Progress in Colloid and Polymer Science
Volume 139

Series Editors
F. Kremer, Leipzig
W. Richtering, Aachen

Progress in Colloid and Polymer Science

Recently published and Forthcoming Volume

UK Colloids 2011
Volume Editors: Victor Starov, Peter Griffiths
Vol. 139, 2012

Trends in Colloid and Interface Science XXIV
Volume Editors: Victor Starov, Karel Procházka
Vol. 138, 2011

Trends in Colloid and Interface Science XXIII
Volume Editor: Seyda Bucak
Vol. 137, 2010

Gels: Structures, Properties, and Functions
Volume Editors: Masayuki Tokita, Katsuyoshi Nishinari
Vol. 136, 2009

Colloids for Nano- and Biotechnology
Volume Editors: Hórvölgyi, Z.D., Kiss, É.
Vol. 135, 2008

Surface and Interfacial Forces From Fnndamentals to Applications
Volume Editors: Auernhammer, G.K., Butt, H.-J , Vollmer, D.
Vol. 134, 2008

Smart Colloidal Materials
Volume Editor: Richtering, W.
Vol. 133, 2006

Characterization of Polymer Surfaces and Thin Films
Volume Editors: Grundke, K., Stanun, M., Adler, H.-J.
Vol. 132, 2006

Analytical Ultracentrifugation VIII
Volume Editors: Wandrey, C., Cölfen, H.
Vol. 131, 2006

Scattering Methods and the Properties of Polymer Materials
Volume Editors: Stribeck, N., Smarsly, B.
Vol. 130, 2005

Mesophases, Polymers, and Particles
Volume Editors: Lagaly, G., Richtering, W.
Vol. 129, 2004

Surface and Colloid Science
Volume Editor: Galembeck, F.
Vol. 128, 2004

Analytical Ultracentrifugation VII
Volume Editors: Lechner, M.D., Börger, L.
Vol. 127, 2004

Trends in Colloid and Interface Science XVII
Volume Editors: Cabuil, V., Levitz, P., Treiner, C.
Vol. 126, 2004

From Colloids to Nanotechnology
Volume Editors: Zrinyi, M., Horvolgyi, Z.D.
Vol. 125, 2004

Aqueous Polymer Dispersions
Volume Editor: Tauer, K.
Vol. 124, 2004

Trends in Colloid and Interface Science XVI
Volume Editors: Miguel, M., Burrows, H.D.
Vol. 123, 2004

Aqueous Polymer – Cosolute Systems
Volume Editor: Anghel, D.F.
Vol. 122, 2002

Victor Starov · Peter Griffiths
Editors

UK Colloids 2011

An International Colloid and Surface Science Symposium

Springer

Editors
Victor Starov
Loughborough University
Chemical Engineering
Loughborough
UK

Peter Griffiths
School of Chemistry
Cardiff University
Cardiff
UK

ISSN 0340-255X ISSN 1437-8027 (electronic)
ISBN 978-3-642-28973-6 ISBN 978-3-642-28974-3 (eBook)
DOI 10.1007/978-3-642-28974-3
Springer Heidelberg New York Dordrecht London

Library of Congress Control Number: 2012943935

Printed on acid-free paper

Springer is part of Springer Science+Business Media (www.springer.com)

Preface

"Colloids 2011" – the first multi-day conference on the topic of colloid science held in the UK for many years – took place in the purpose-built De Vere conference centre in the heart of Canary Wharf, London, from July 4th to 6th 2011. Jointly organized by the RSC Colloid and Interface Science Group and the SCI Colloid and Surface Science Group, it was designed as the perfect opportunity for UK and international researchers interested in colloids and interfaces to meet, present and discuss issues related to current developments in this field. Canary Wharf is the centre of London's business district and the many bars and restaurants within the vicinity of the conference venue made for an equally enjoyable social setting.

The conference had over 250 delegates, from all across the world – good representation from Japan, China, Australia, USA, France, Germany, Holland, Sweden, Spain, Poland, Georgia – as well as a substantial number of UK based researchers. Lectures, organised into three parallel sessions, numbered 100 ordinary presentations, 12 Keynote and 4 Plenary lectures. The conference also hosted the new Thomas Graham Lecture, created to recognise a scientist who has already made a distinguished contribution to the field of colloid science, with the prospect of a further 15+ years of active research to come. The first ever Thomas Graham Lecture was awarded to Prof Colin Bain (University of Durham) for his outstanding work on the characterisation and properties of interfaces and adsorbed layers. The conference also had about 75 poster presentations that ran in conjunction with a successful exhibition.

This Special Issue of "Progress in Colloid and Polymer Science" collects together a selection of 20 papers mostly presented during the Conference. The papers included cover the wide variety of topics from fundamentals in colloid and interface science to industrial applications. The current Special Issue reflects also an international character of the Conference.

Loughborough, UK — Victor Starov
Cardiff, UK — Peter Griffiths

Contents

Progr Colloid Polym Sci (2012) 139
DOI: 10.1007/978-3-642-28974-3_1

Spreading and Evaporation of Surfactant Solution Droplets

Hezekiah Agogo[1], Sergey Semenov[2], Francisco Ortega[1], Ramón G. Rubio[1], Víctor M. Starov[2], and Manuel G. Velarde[3]

Abstract Evaporation of liquid droplets in gas volume has implications in different areas: spray drying and production of fine powders [1–3], spray cooling, fuel preparation, air humidifying, heat exchangers, drying in evaporation chambers of air conditioning systems, fire extinguishing, fuel spray auto ignition (Diesel), solid surface templates from evaporation of nanofluid drops (coffee-ring effect), spraying of pesticides[1–4], painting, coating and inkjet printing, printed MEMS devices, micro lens manufacturing, spotting of DNA microarray data [3–5]. Because of such wide range of industrial applications this phenomenon has been under investigation for many years, both in the case of pure and multicomponent fluids. The studies encompass different conditions: constant pressure and temperature, elevated pressure, fast compression, still gas atmosphere and turbulent reacting flows, strongly and weakly pinning substrates [1, 2]. Even though experimental, theoretical and computer simulation studies have been carried out [1–11], and have taken into account different physical processes; heat transfer inside droplets, mass diffusion in bi- and multi- component fluids, droplet interactions in sprays, turbulence, radiation adsorption, thermal conductivity of the solid substrate, Marangoni convection inside the droplets.

Introduction

Evaporation of liquid droplets in gas volume has implications in different areas: spray drying and production of fine powders [1–3], spray cooling, fuel preparation, air humidifying, heat exchangers, drying in evaporation chambers of air conditioning systems, fire extinguishing, fuel spray auto ignition (Diesel), solid surface templates from evaporation of nanofluid drops (coffee-ring effect), spraying of pesticides [1–4], painting, coating and inkjet printing, printed MEMS devices, micro lens manufacturing, spotting of DNA microarray data [3–5]. Because of such wide range of industrial applications this phenomenon has been under investigation for many years, both in the case of pure and multicomponent fluids. The studies encompass different conditions: constant pressure and temperature, elevated pressure, fast compression, still gas atmosphere and turbulent reacting flows, strongly and weakly pinning substrates [1, 2]. Even though experimental, theoretical and computer simulation studies have been carried out [1–11], and have taken into account different physical processes; heat transfer inside droplets, mass diffusion in bi- and multi- component fluids, droplet interactions in sprays, turbulence, radiation adsorption, thermal conductivity of the solid substrate, Marangoni convection inside the droplets.

Cioulachjian et al. [4] investigated the drop evaporation phenomena under moist air or saturated conditions using water as the pure liquid, they found an influence from surroundings on droplet evaporation dynamics. Shanahan and Bourges-Monnier [2] used large drops of water and *n*-decane on polyethylene, epoxy resin, and Teflon, both in a saturated vapor atmosphere and in open air. They showed the existence of four distinct stages in the evaporation process in open air conditions. Sefiane et al. [3] obtained results which indicate that evaporation rate is significantly reduced as the relative humidity is increased, while the initial values of θ and L remain constant, with ethanol/water mixture from mildly diluted to highly concentrated solutions they showed a stage of the process although a different evaporation pattern due to the volatility of one of the components here. Doganci et al. [6] carried out evaporation experiments using SDS at various concentrations above and below the CMC, they found that an increase in surfactant concentration reduces the interfacial tension between droplet and the hydrophobic substrate significantly and that evaporation occurs

R.G. Rubio (✉)
[1]Departamento de Química Física I, Universidad Complutense, Madrid, 28040, Spain
e-mail: rgrubio@quim.ucm.es
[2]Department of Chemical Engineering, Loughborough University, Loughborough, UK
[3]Unidad de Fluidos, Instituto Pluridisciplinar, Universidad Complutense, Madrid, 28040, Spain

by diffusion because the rate of change of the droplet volume raised to the power 2/3 ($V^{2/3}$) with time remained linear throughout the experiment. Sultan et al. [7] derived sets of equations which describe the evaporation of a thin layer. They applied the lubrication approximation combined with the diffusion limited evaporation with the thermodynamic rate of transfer across the liquid–gas interface and were able to obtain a correlation between the instability of a uniform liquid layer and the festoon instability of an evaporating droplet. Rednikov et al. [8] investigated the microstructure of a contact line formed by a liquid and its pure vapor for a completely wetted superheated substrate with a disjoining pressure in the form of a positive inverse cubic law. From the results it can be inferred that the regime of a truncated liquid microfilm on a solid surface can be thermodynamically more stable than the regime of an extended microfilm.

For the spreading of a thin volatile liquid droplet on a uniformly heated surface a mathematical model was developed by Ajeav [9], the lubrication-type approximation and one sided model, when density, dynamic viscosity and heat conductivity of gas are small compared to the liquid, they found that the dynamics of a droplet is a competition between evaporation and spreading. Moroi et al. [10] compared the evaporation rate of water and SDS solutions across air/surfactant solution interface using a thermo-gravimetric balance they found no apparent difference between the evaporation rate and activation energy for surfactant solutions below and above the CMC.

Some phenomenological results are now well established for the evaporation of a drop of a multicomponent fluid onto a perfectly smooth surface under partial wetting conditions:

(a) Assuming the evaporation process right after the drop has spread over the solid substrate, and that the drop has a spherical shape, the evaporation rate is proportional to the radius of the drop onto the substrate, L.
(b) The evaporation process is composed of four stages: (1) L increases while the contact angle, θ, decreases down to the advancing contact angle value, θ_{ad}. (2) The contact angle decreases from θ_{ad} down to receding contact angle value, θ_{red}, at constant L. (3) Contact angle remains constant and equal its receding value θ_r, while the radius of the base droplet, L, decreases. (4) Both the contact angle and L decrease until the drop completely evaporates. For the purpose of understanding the evaporation stages we will discuss the second and third stages.

Nevertheless many problems still remain to be solved: (a) To build a theory for drops multicomponent fluids that include all the physical processes above mentioned. (b) To build a hydrodynamic model able to describe the four stages of the evaporation process. (c) To compare the hydrodynamic description of the drop evaporation with the molecular thick layer beyond the three-phase contact line. Such comparison must take into consideration the DLVO forces acting at a mesoscopic scale near the contact line. (d) To describe the evaporation process of complex fluids: polymer and protein solutions and nanoparticle suspensions. (e) To describe the evaporation of drops onto patterned surfaces.

In a recent work we have carried out computer simulations of the evaporation of a drop of pure fluid. The results have shown that evaporation process of small sessile water droplets in the presence of contact angle hysteresis can be subdivided into four stages. Introducing a dimensionless contact line radius and dimensionless times for each stage allowed us to deduce universal laws describing two stages of the evaporation process in the case of contact angle hysteresis. The theory describes two stages and shows good agreement with experimental data available in the literature [11]. Partly based on these results a model was proposed that was able to quantitatively explain two stages of the evaporation of pure fluids. In fact, the model has allowed us to build universal curves of the time dependence of contact angle and droplet base radius, L.

The aim of this work is to perform a detailed experimental study of the time dependence of the contact angle, the volume and the radius of the drop onto a hydrophobic TEFLON-AF substrate. We have used drops of an aqueous solution of a super spreader surfactant (Silwett 77) over all concentration ranges i.e. below and above the critical aggregation concentration (C.A.C).

Experimental Technique

SILWET L77 was purchased from Sigma-Aldrich (Germany) and used as received. Poly (4, 5-difluoro-2, 2-bis (trifluorimethyl)-1, 3-dioxole-co-tetrafluoroethylene), denoted TEFLON-AF, was purchased from Sigma-Aldrich (Germany) as powder, the Fuorinet F-77 solvent was bought from 3 M (U.S.A.), and the silicon wafers were obtained from Siltronix (France). Ultrapure deionized water (Younglin Ultra 370 Series, Korea) with a resistivity higher than 18 MΩ and TOC lower than 4 ppm.

All the surfactant solutions were prepared by weight using a balance precise to ± 0.01 mg. A pH=7.0 buffer was used as solvent to prevent hydrolysis of the SILWET L77. The solutions were used immediately after preparation. The silicon wafers were cleaned using piranha solution for 20 min (caution piranha solution is highly oxidizing!) The solid substrates were prepared as follows: the TEFLON-AF powder was suspended in the Fuorinet F77 and spin-coated onto the silicon wafers. The average roughness of the 20×20 μm surface was ≈ 1.0 nm as measured by AFM (tapping mode). The macroscopic contact angle of pure water was (104 ± 2) on those substrates. Drops of 4 mm^3 were deposited onto the substrate for measurements. Five independent measurements were done for each experimental point reported in this work, and only those that agreed within the experimental uncertainty were accepted.

The experimental technique used was similar to the one used in the papers by Ivanova et al. [12, 13], with some modifications that allowed us to monitor continuously the temperature and the relative humidity inside the experimental set up. Figure 1 shows a scheme of the device.

Sessile drops are deposited onto the substrate inside a chamber attached to a thermostat, and its shape and size was captured by one CCD camera (side view) at a rate of 30 fps. The drop volume is maintained at about 4 mm^3 in order to ensure that gravity effects can be neglected and the drop always has a spherical cap shape. The images captured were analyzed using the drop tracking and evaluation analysis software (Micropore Technologies, U.K.) that allowed us to monitor the time evolution of the drop base diameter, height, radius of curvature, and contact angle of the drops. The precision of the contact angle was ±2 under dynamic conditions, i.e. spreading and evaporation, and that of the temperature was ±1°C The relative humidity, was managed by placing a saturated salt solution inside the measuring chamber (KBr and KCl for a relative humidity approx. 90%) and it was measured with a precision of ±2%.

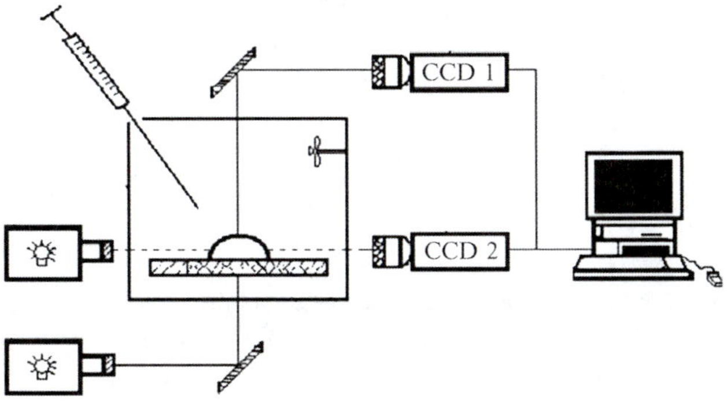

Fig. 1 Scheme of the experimental technique. CCD1 and CCD2 are the cameras to capture the drop profiles from the top and from the side. Inside the camber both the temperature and the relative humidity are controlled and continuously monitored

Results and Discussion

Figure 2 shows typical results of water and Silwet L-77 solutions at different concentrations, the increase of surfactant concentration reduces the initial contact angle and increases the initial perimeter of the droplet

It is worth mentioning that the rate at which the droplet spreads before the radius reaches a maximum has been found to be proportional to the surfactant concentration.

Figure 3 shows an evaporation profile for Silwet L-77 at CMC with the four stage evaporation profile which is being investigated. The data is analyzed until the errors in measurements become unreasonably high due to the small initial volumes. We will analyze the influence or temperature and relative humidity on evaporation rate in subsequent publications.

The volume of the droplet to power 2/3 decreases linearly with time in our experiments from which can infer that evaporation is diffusion controlled. We observe that the droplet evaporation profile in terms of contact angle and droplet radius dependencies with respect to time is consistent with the droplet volume linear dependency on time (Fig. 4).

Theoretical Background

Figure 5 illustrates the geometry of the problem of spherical cap droplet evaporation (we assume that the spreading process has already ended).

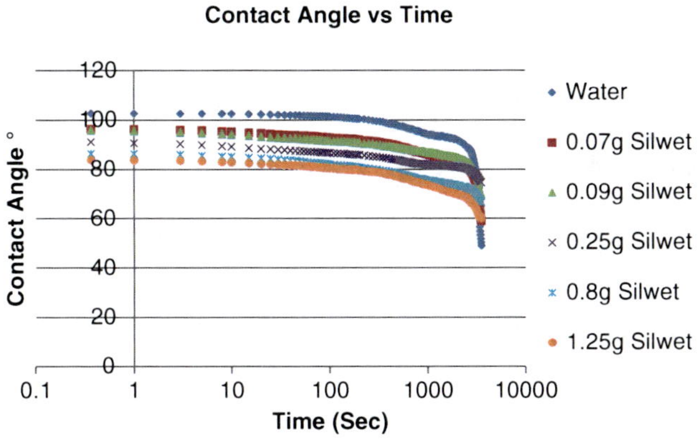

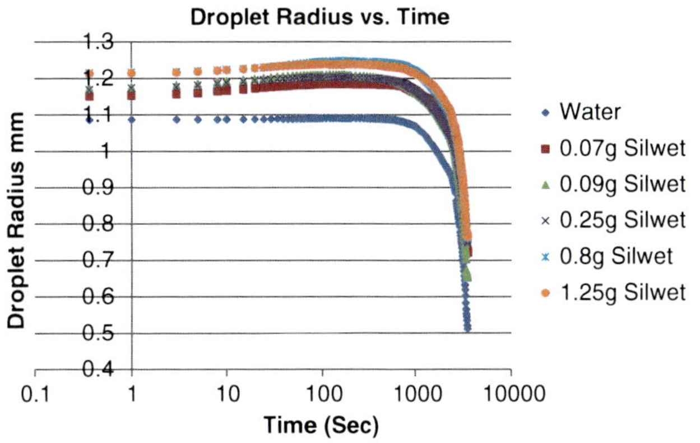

Fig. 2 Time dependence of the contact angle and of the radius of the droplet base for water and different concentrations of surfactant at 18°C and 90% relative humidity

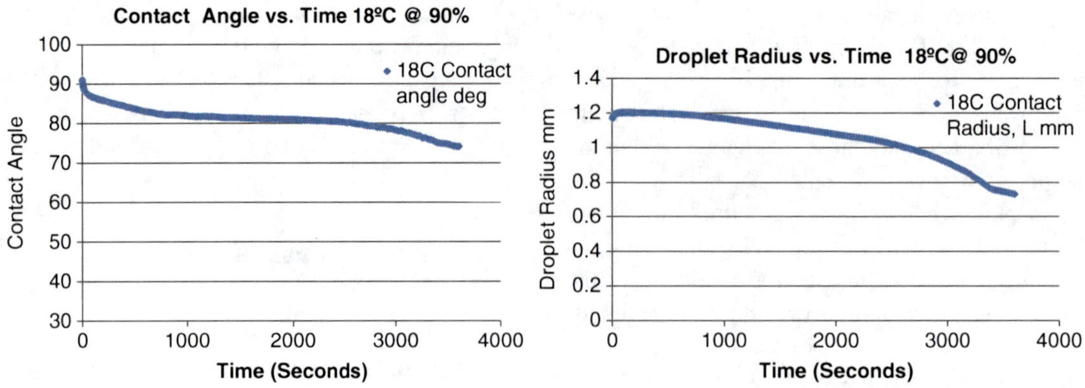

Fig. 3 Temperature effect on the time dependence of the contact angle and of the contact radius for a surfactant concentration 0.25 CAC at 18°C and 90% relative humidity

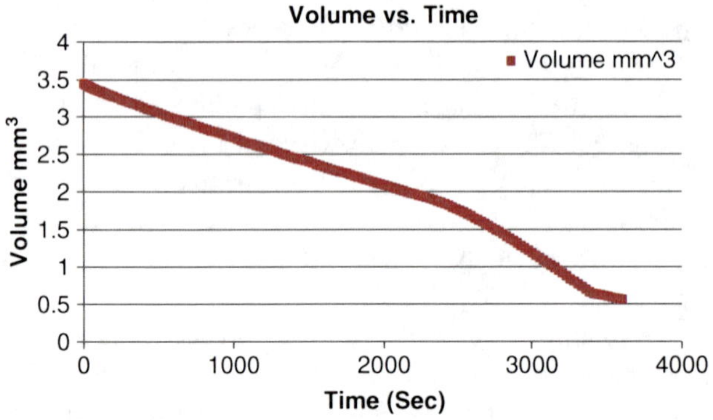

Fig. 4 Volume dependence on time with respect to humidity and temperature at 18°C and 90% for surfactant concentrations 0.25CAC

In recent work, Semenov et al. [11, 14] have shown that if during the two first stages of evaporation the drop remains spherical; its volume is given by:

$$V = L^3 f(\theta)\, f(\theta) = \frac{\pi}{3}\frac{(1-\cos\theta)^2(2+\cos\theta)}{\sin^3\theta} \tag{1}$$

On the other hand, computer simulations in [11] have shown that:

$$\frac{dV}{dt} = -\beta F(\theta) L \tag{2}$$

where

$$F(\theta) = \begin{cases} (0.6366\cdot\theta + 0.09591\cdot\theta^2 - 0.06144\cdot\theta^3)/\sin\theta, \\ \theta<\pi/18 \\ (0.00008957 + 0.6333\cdot\theta + 0.116\cdot\theta^2 \\ -0.08878\cdot\theta^3 + 0.01033\cdot\theta^4)/\sin\theta, \\ \theta \geq \pi/18 \end{cases} \tag{3}$$

$$\beta = 2\pi\frac{DM}{\rho}[c_{sat}(T_{av}) - c_\infty] \tag{4}$$

and $T_{av} = \int_S TdS$ is the average temperature of the droplet surface, S. In (4) D is the diffusion coefficient of the liquid vapor in air, ρ is the liquid density and M its molecular weight. The computer simulations suggest that, for given values of temperature, relative humidity and surfactant concentration, T_{av} and β can be taken as constants during the evaporation process.

During the first stage of evaporation L remains constant and equal to its initial value L_o. Then (2) can be rewritten as

$$L_0^2 \cdot f'(\theta)\frac{d\theta}{dt} = -\beta \cdot F(\theta) \tag{5}$$

From which time dependence of the contact angle can be easily obtained after numerical integration. The theory predicts a universal behavior for this evaporation stage when the variables are expressed in terms of a reduced time, $\tilde{\tau}$, defined as

$$\tilde{\tau} = \tau + \int_{\theta_{ad}}^{\pi/2} f'(\theta)/F(\theta)\, d\theta, \qquad \tau = t/t_{ch}, \qquad t_{ch} = L_0^2/\beta. \tag{6}$$

For the second evaporation stage, where θ is constant and equals its receding value θ_r, the theory also predicts $l(\theta) = \left[1 - \frac{2F(\theta_r)}{3f(\theta_r)}(\tau - \tau_r)\right]^{1/2}$, where $l = L/L_0$ is a reduced radius of the three-phase contact line, and reduced time τ_r corresponds to the moment when receding starts. That represents a universal behavior for the time dependence of the reduced radius of the three-phase contact line, l, on reduced time. In terms of new reduced time $\bar{\tau} = \frac{2F(\theta_r)}{3f(\theta_r)}(\tau - \tau_r)$ it takes the following form: $l(\bar{\tau}) = (1-\bar{\tau})^{1/2}$.

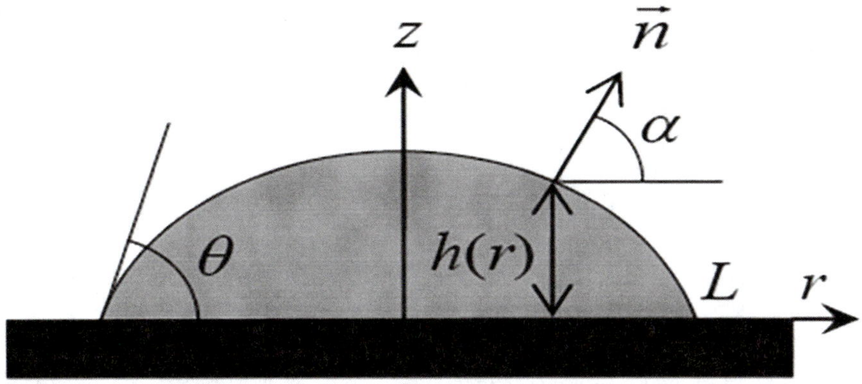

Fig. 5 Geometry of the problem of a spherical cap droplet that evaporates onto a solid (partial wetting conditions). L is the radius of the basis of the droplet, θ is the contact angle, c_{sat} and c_{∞} are the solvent concentrations at the surface and in the vapor phase far from the droplet surface

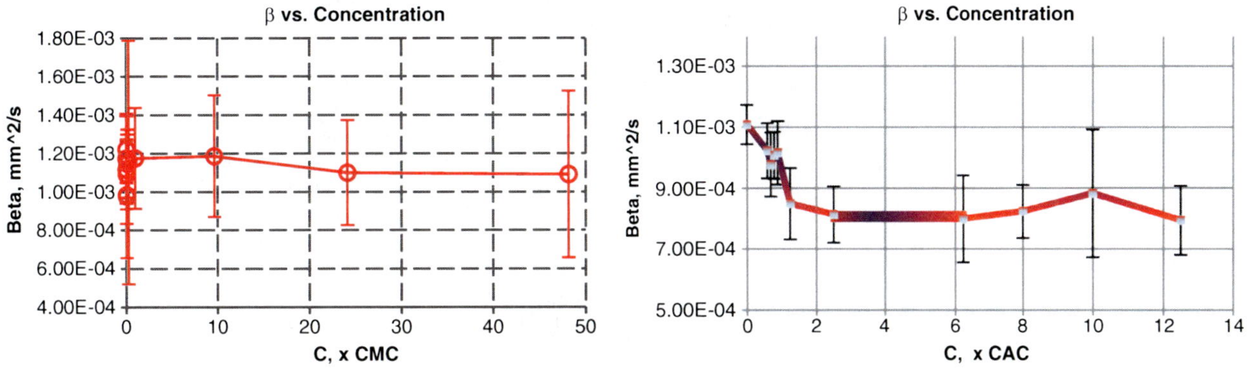

Fig. 6 Surfactant concentration dependence of the β parameter for SDS and Silwet L-77 Solutions at similar experimental conditions (55% RH, 21.7°C for SDS and 90% RH, 18°C for Silwet L-77 Surfactants)

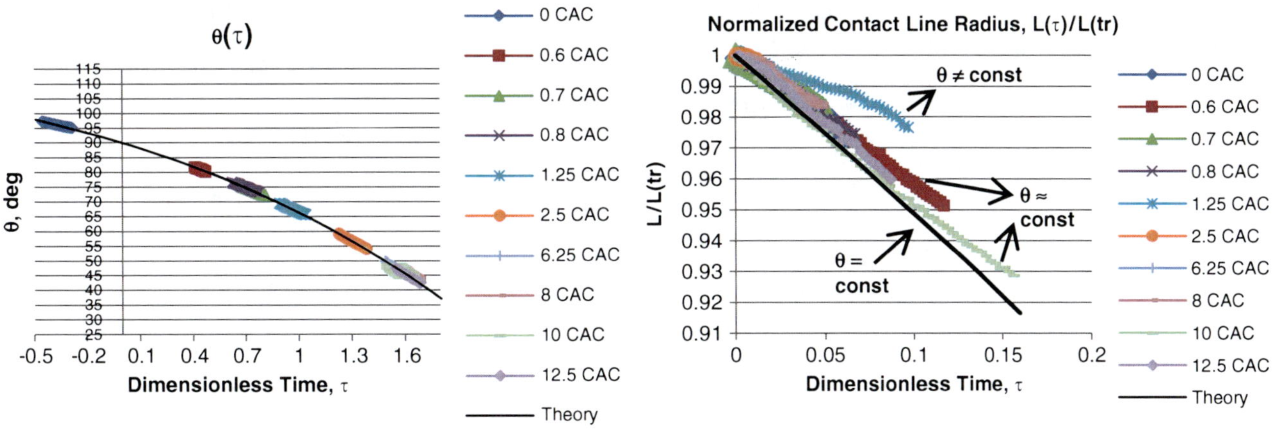

Fig. 7 Comparison of the universal behavior predicted by the theory and the experimental results of surfactant solutions

The theoretical predictions were found to predict reasonably well the available data for water droplets onto different solid substrates. In what follows the theoretical predictions will be compared with the experimental results for surfactant solutions.

Validation Against Experimental Data

We show the behavior of the developed parameter β as a function of SILWET L77 concentration. Preliminary results are shown in Fig. 6 where the predictions of the universal behaviors are tested based on our experiments using Silwett L-77. It can be observed that during the first stage of the evaporation process the agreement between theory and experiment is very good, furthermore, changing the surfactant concentration has allowed us to test a very broad range of the reduced time.

Comparing some of the data obtained by Doganci et al. in their experiments using SDS surfactant (55% RH, 21°C) with our experiments using SilwetL-77 (90% RH, 18°C) we have the following plots one concludes that both systems have similar behaviors.

Within the limits of experimental errors it may be correct to infer that β remains constant at all concentrations which further agrees with the data from Doganci et al. from their experiments that surfactant concentrations does not influence the evaporation rate, it does however reduce the surface tension at the liquid –solid interface which leads to an increase in the spreading potential proportionally (Fig. 7).

The experimental data follow the predictions only where the transition from the first stage of evaporation to the second stage can be determined; the existence of a third stage which is not modeled into this theory is still being analyzed.

Conclusion

Increasing the surfactant concentration in solution the advancing contact angle reduces from 104 ± 2 until 83 ± 2 which is explained by the Teflon surface becoming increasingly hydrophilic. Surfactant molecules are being adsorbed at the water-air interface the interfacial tension between the droplet and the substrate. All volume ($V^{2/3}$) plots as a function of time are linear which allows us to infer that evaporation is mainly by diffusion. The contact angle of sessile droplets decrease with increasing surfactant concentration however it has no considerable effect on the rate of evaporation which is in agreement with other literature.

References

1. Semenov S, Starov VM, Velarde MG, Rubio RG (2011) Droplets evaporation: problems and solutions, European physical journal special topics. Eur Phys J Spec Top 197:265–278
2. Bourges-Monnier C, Shanahan M (2005) Influence of evaporation on contact angle. Langmuir 11:2820–2829
3. Sefiane K, Tadrist L, Douglas M (2003) Int J Heat Mass Trans 46:4527–4534
4. Hua Hu, Ronald G (2002) Larson. J Phys Chem B 106:1334–1344
5. Cioulachjian S, Launay S, Boddaert S, vLallemand M (2010) Int J Therm Sci 49:859–866
6. Doganci MD, Sesli BU, Erbil HY (2011) Diffusion-controlled evaporation of sodium dodecyl sulfate solutions drops placed on a hydrophobic substrate. J Colloid Interf Sci 362:524–531
7. Sultan E, Boudaoud A, Amar MB (2005) Evaporation of a thin film: diffusion of the vapour and marangoni instabilities. J Fluid Mech 543:183–202
8. Ye A, Rednikov P (2011) Colinet, truncated versus extended microfilms at a vapor-liquid contact line on a heated substrate. Langmuir 27(5):1758–1769
9. Ajaev VS (2005) Spreading of thin volatile liquid droplets on uniformly heated surfaces. J Fluid Mech 528:279–296
10. Moroi Y, Rusdi M, Kubo I (2004) J Phys Chem B 108:6351–6358
11. Semenov S, Starov VM, Rubio RG, Agogo H, Velarde MG (2011) Evaporation of sessile water droplets: universal behavior in presence of contact angle hysteresis. Colloid Surf. A Physicochem Eng Asp 391:135–144
12. Ritacco H, Ortega F, Rubio RG, Ivanova N, Starov VM (2010) Equilibrium and dynamic surface properties of trisiloxane aqueous solutions part 1 experimental results. Colloids Surf A Physicochem Eng Asp 365:199–203
13. Ivanova N, Starov V, Johnson D, Hilal N, Rubio RG (2009) Spreading of aqueous solutions of trisiloxanes and conventional surfactants over PTFE AF coated silicon wafers. Langmuir 25:3564–3570
14. Semenov S, Starov VM, Rubio RG, Velarde MG (2010) Instantaneous distribution of fluxes in the course of evaporation of sessile liquid droplets: computer simulations. Colloids Surf A Physicochem Eng Asp 372:127–134

Progr Colloid Polym Sci (2012) 139
DOI: 10.1007/978-3-642-28974-3_2

Novel Membrane Emulsification Method of Producing Highly Uniform Silica Particles Using Inexpensive Silica Sources

Marijana M. Dragosavac[1], Goran T. Vladisavljević[1,2], Richard G. Holdich[1,3], and Michael T. Stillwell[3]

Abstract A membrane emulsification method for production of monodispersed silica-based ion exchange particles through water-in-oil emulsion route is developed. A hydrophobic microsieve membrane with 15 μm pore size and 200 μm pore spacing was used to produce droplets, with a mean size between 65 and 240 μm containing acidified sodium silicate solution (with 4 and 6 wt% SiO_2) in kerosene. After drying, the final silica particles had a mean size in the range between 30 and 70 μm. Coefficient of variation for both the droplets and particles did not exceed 35%. The most uniform particles had a mean diameter of 40 μm and coefficient of variation of 17%. The particles were functionalised with 3-aminopropyltrimethoxysilane and used for chemisorption of Cu(II) from an aqueous solution of $CuSO_4$ in a continuous flow stirred cell with slotted pore microfiltration membrane. Functionalised silica particles showed a higher binding affinity toward Cu(II) than non-treated silica particles.

Introduction

There has been an increasing interest in the production of functionalised porous silica microspheres for use in analytical, preparative, and ion exchange columns, requiring particle diameters greater than 1 μm. A control of particle size and internal microstructure of silica particles is critically important in ion exchange, biochemical sensing [1], drug delivery [2], catalysis [3], and chromatography. Silica particles can be fabricated from organic silicon precursors, e.g. siliciumalkoxide [4] or inorganic materials, such as silicate solutions [5]. In the latter case, acidified sodium silicate solution is dispersed in an organic phase to form water-in-oil (W/O) emulsion [5] or atomized in air [6]. Silicic acid formed by acidification of sodium silicate immediately undergoes spontaneous condensation polymerisation which progresses within the dispersed droplets, eventually leading to the formation of branched polymer with –Si–O–Si– bonds. Upon drying the formed silica hydrogel shrinks to xerogel. A surfactant can be dissolved in the silica solution to additionally tailor the internal gel structure (surfactant templating) [7]. This paper reports a novel W/O emulsion route based on using microsieve-type membrane to prepare porous silica microspheres with a controllable size between 30 and 70 μm. To the best of our knowledge, membrane emulsification has only been used for fabrication of silica particles up to 3 μm in size and only inorganic membranes were used in this application [8–11].

After functionalisation, the produced silica particles have been used as ion-exchange particles for removal of copper from an aqueous solution. Copper(II) is often used as a model cation for investigation of performance of ion exchangers. In addition, copper is widely present in industrial wastewaters and its disposal represents a big environmental threat, since it can induce severe health problems. Classical techniques for removal of copper from waste waters are flotation, chemical precipitation, and ion exchange with organic resins, such as sulfonated poly(styrene-co-divinylbenzene) resins [12]. In this work, new inorganic ion-exchange particles have been fabricated and tested for copper removal.

Experimental

Preparation of W/O Emulsion

The dispersed phase was prepared by dripping sodium silicate solution, containing 10 or 15 wt% SiO_2, into 1 M H_2SO_4

G.T. Vladisavljević (✉)
[1]Chemical Engineering Department, Loughborough University, Loughborough, Leicestershire LE11 3TU, UK
[2]Vinča Institute of Nuclear Sciences, University of Belgrade, 522, Belgrade, Serbia
e-mail: g.vladisavljevic@lboro.ac.uk
[3]Micropore Technologies Ltd, Hatton, Derbyshire, DE65 5DU, UK

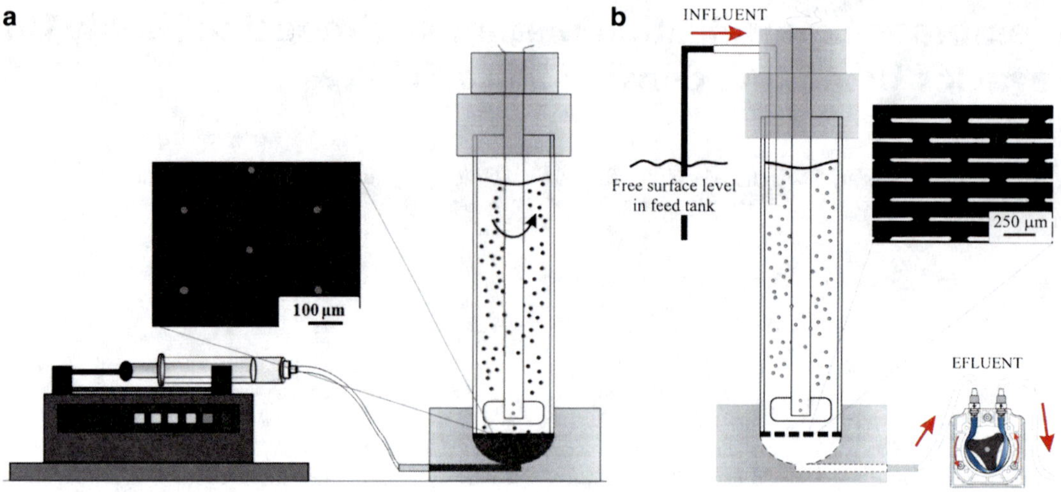

Fig. 1 Schematic illustration of: (**a**) Dispersion Cell containing a paddle stirrer above a disk membrane with cylindrical 15 μm pores used for membrane emulsification and (**b**) microfiltration system with slotted pore membrane used for combined microfiltration/ion exchange experiments

under vigorous stirring until pH of 3.5 was achieved. The continuous phase was 5 wt% Span 80 in low odour kerosene (both supplied by Sigma Aldrich, UK). Emulsification was performed using a Micropore Technologies Ltd. Dispersion Cell (Fig. 1a) equipped with a hydrophobic nickel membrane with 15 μm pore size, 200 μm spacing between the pores and 8.5 cm^2 effective membrane area. The cell was filled with 100 ml of continuous phase and 10 ml of dispersed phase was injected at constant controllable rate using a Harvard Apparatus model 11 Plus syringe pump.

Production of Silica Particles

The resultant W/O emulsion was transferred to a beaker, diluted with kerosene and continuously stirred until the hydrogel was formed. Beside the spherical silica particles, needle shapes of silica were also created. To filter them out, a microfiltration cell with slotted pore membrane was used. The membrane with an effective membrane area of 8.5 cm^2 was provided by Micropore Technologies Ltd. Slots were 4 μm wide and 250 μm long enabling that even the smallest silica particles were maintained above the membrane. No spherical particles were observed in the filtrate and the amount of needle-like silica was less than 2% of the total particle mass being filtered. After removal of non-spherical silica, the particles were washed with at least 250 ml of acetone followed by water washing to remove Span 80 and kerosene. After that, the particles were dried at room temperature and calcined at 550°C with a ramp step of 20°C min^{-1} for 6 h. Specific surface area of calcined silica particles was determined by the adsorption of nitrogen gas using Micromeritics ASAP 2020 [13].

Functionalisation of Calcined Silica Particles and Combined Microfiltration and Ion Exchange

Ten gram of calcined silica particles, with a specific surface area of 360 m^2 g^{-1} was first washed with a mixture of nitric and hydrochloric acid in a ratio of 1:3 for 2 h in order to remove possible metal impurities. The particles were then filtered, dried in a vacuum at 200°C, and refluxed in a mixture of 80 ml of toluene and 10 ml of 3-aminopropyltrimethoxysilane for 24 h. After that, the particles were collected by filtration, washed with ethanol and then transferred to a Soxhlet extractor and washed with toluene for 24 h, to eliminate possible traces of 3-aminopropyltrimethoxysilane. According to Lam et al. [14] silica functionalised with 3-aminopropyltrimethoxysilane is capable of adsorbing copper ions from solutions. Combined microfiltration and ion exchange was carried out at room temperature in a continuous flow stirred cell (Fig. 1b) provided by Micropore Technologies Ltd. A metal membrane with 8 × 400 μm slots was fitted to the bottom of the cell. This slot width was fine enough to keep all particles in the cell. An aqueous solution of $CuSO_4$ containing 10 gcm^{-3} Cu(II) was continuously delivered to the cell at a constant flow rate of 8 ml min^{-1}. Ten milliliter samples of the effluent stream were taken at regular time intervals to determine Cu(II) concentration using an Atomic Absorbance Spectrophotometer (Spectra AA-200 Varian).

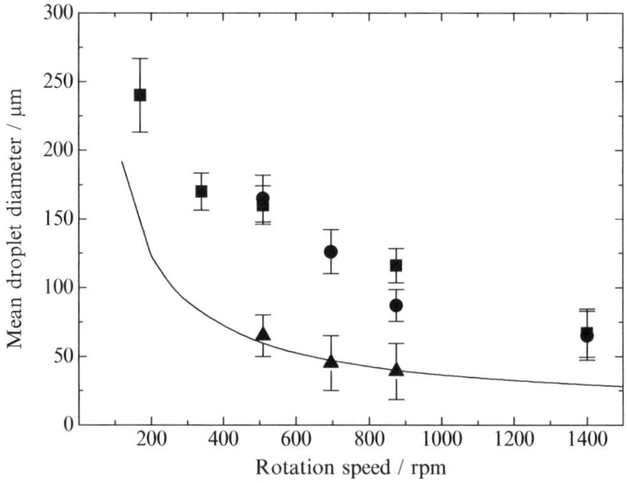

Fig. 2 Droplet diameters of produced W/O emulsions as a function of rotation speed. Line represents the model prediction valid at low injection rates [15]. Dispersed phase: (▪) Sodium silicate with 6 wt% SiO_2 in 1 M H_2SO_4 injected at 350 $Lm^{-2}\,h^{-1}$. (•) Sodium silicate with 4 wt% SiO_2 in 1 M H_2SO_4 injected at 350 $Lm^{-2}\,h^{-1}$. (▲) Sodium silicate with 6 wt% SiO_2 in distilled water injected at 1 $Lm^{-2}\,h^{-1}$. Error bars in the figure represent one standard deviation ($\sigma = CV \times d/100$) from the mean value

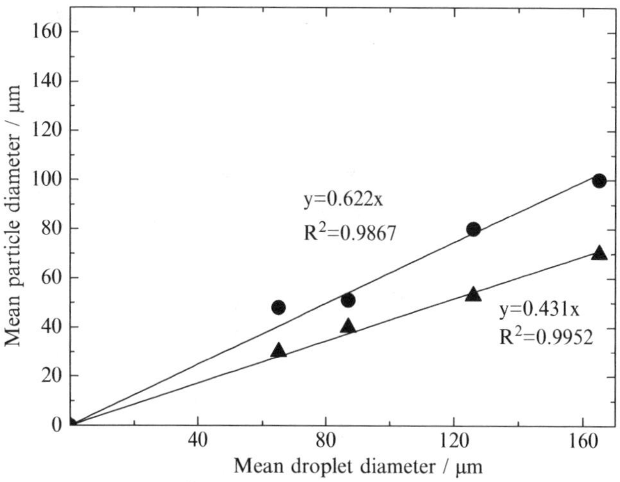

Fig. 3 Relationship between the particle diameter in the resultant gels and the droplet diameter. Dispersed phase: sodium silicate with 6wt% SiO_2 in 1 M H_2SO_4 injected at 350 $Lm^{-2}\,h^{-1}$. (•) Hydrogel. (▲) Xerogel

Production of Silica Gel Using Membrane Emulsification

Figure 2 illustrates the effect of paddle rotation speed and flow rate of the dispersed phase on the droplet size of the W/O emulsions produced. An increase of the rotation speed increases the shear stress on the membrane surface and the droplet formation time shortens, therefore, the smaller droplets are produced. The force balance model based on average shear stress [15] provided a good prediction of the droplet size for a very low flux of 1 $Lm^{-2}\,h^{-1}$. The parameters used for modeling are provided elsewhere [15]. Droplets produced using the higher injection rate are larger than the ones produced at lower injection rate and do not fit the model curve. Two explanations of this phenomenon are possible. First, the model does not take into consideration a dispersed phase inflow during the necking time. The detachment of the droplet is not instantaneous but requires a finite time, the necking time [16], during which an additional amount of dispersed phase flows into the droplet. Secondly, the model calculations are made using the equilibrium interfacial tension (2.7 $mN\,m^{-1}$) which is lower than the actual interfacial tension during drop formation. It should be noted that the interfacial tension if no surfactant is present is 16 $mN\,m^{-1}$ and during drop generation the liquid/liquid interface is not fully saturated with surfactant molecules. The produced droplets were stirred in a beaker in order to allow gelling and formation of a hydrogel. Once gelled, the hydrogel particles were dried to form a xerogel. Figure 3 illustrates the shrinkage of the dispersed phase droplets during their transformation into silica gel particles. During condensation polymerization the droplets will shrink due to water loss as the hydrogel is formed. The hydrogel particles were left to dry for several days at room temperature. During this drying stage liquid present in the pores is removed, the structure compresses and the porosity is reduced by the surface tension forces as the liquid is removed leaving dried silica particles (xerogel). Drying at room temperature was followed by calcination, but further shrinkage of the dried silica particles was not observed. The size of final silica particles was found to be 2.3 times smaller than the initial droplet size. However, the particles are still significantly bigger than would be predicted by a mass balance of the silica used in their formation. It can be explained by a significant amount of internal porosity, which is 88% and 64% for the hydrogel and xerogel particles, respectively. The coefficient of variation (CV) for xerogel particles was below 31% and was minimal (17%) for the particle diameter of 40 μm.

The surface structure of the silica particles after calcination was imaged by SEM (operated at 2.6 kV) and FEG SEM (operated at 10 kV) and microphotographs are presented in Fig. 4. The silica particles are almost perfectly round as can be seen from Fig. 4a and b, while the close-up of the particle surface shows a cloudy and corrugated external and internal surface morphology (Fig. 4c, d). The presence of pores is visible on the surface.

Functionalisation of Silica Gel and Combined Microfiltration/Ion Exchange

The ability of functionalised particles to adsorb Cu(II) was demonstrated in a continuous flow stirred cell. The volume of the liquid phase in the cell was 140 ml and the stirrer

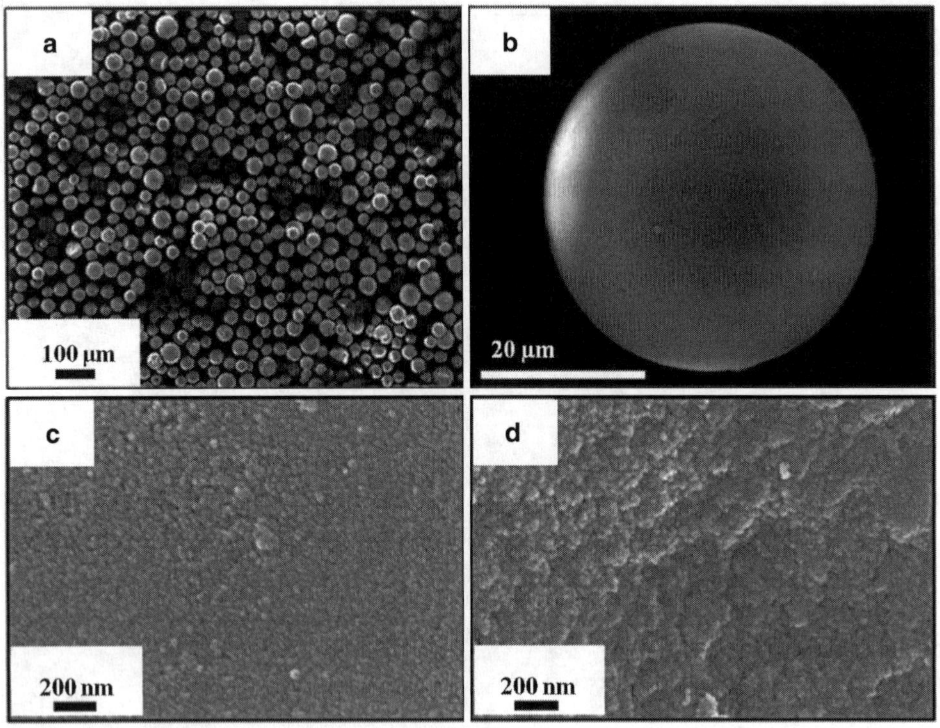

Fig. 4 (**a**) Scanning electron micrograph (SEM) of the silica particles with an average size of 40 μm. (**b**) SEM of a single silica sphere. (**c**) Field emission gun (FEG) SEM of a silica sphere external surface structure. (**d**) FEG SEM of a broken silica sphere

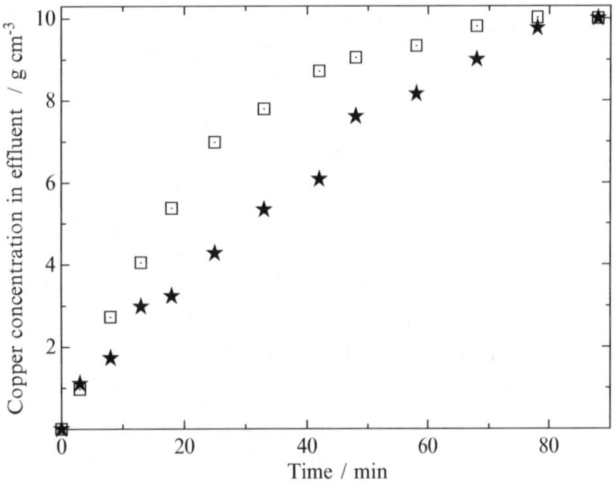

Fig. 5 Chemisorption of $CuSO_4$ on silica gel in a continuous flow stirred cell with slotted pore membrane. The copper concentration in influent was 10 gcm^{-3}, the flow rate of influent was 7 ml min^{-1}, and the particle loading in the cell was 2.5 g. (□) Non-treated silica gel. (★) Silica gel functionalised with 3-aminopropyltrimethoxysilane

speed was 270 rpm. As can be seen from response curves in Fig. 5, the functionalised silica particles had a higher binding affinity toward Cu(II). After about 80 min, both non-treated and functionalized particles were fully saturated with Cu(II) and from that time the Cu(II) concentration in the effluent matched the Cu(II) content in the feed stream.

Conclusions

Spherical silica particles with mean particle sizes controllable within a range between 30 and 70 μm have been produced from inexpensive sodium silicate solution using membrane emulsification. The initial droplet size was precisely controlled by controlling the stirrer speed and the injection rate of the dispersed phase through the membrane and the final particle size was 2.7 times smaller than the initial droplet size. The internal porosity of the particles aged in water was 88% and 64% for the hydrogel and xerogel, respectively. Xerogel particles were functionalised with 3-aminopropyltrimethoxysilane and successfully used as ion-exchange media for chemisorption of Cu(II) from aqueous solutions of $CuSO_4$.

Acknowledgement This research was supported by Engineering and Physical Sciences Research Council, UK (DIAMOND project into Decommissioning, Immobilisation And Management Of Nuclear wastes for Disposal).

References

1. Buranda T, Huang J, Ramarao GV, Linnea K, Larson RS, Ward TL, Sklar LA, Lopez GP (2003) Langmuir 19:1654
2. Barbé C, Bartlett J, Kong L, Finnie K, Lin HQ, Larkin M, Calleja S, Bush A, Calleja G (2004) Adv Mat 16:1959

3. Suzuki TM, Yamamoto M, Fukumoto K, Akimoto Y, Yano K (2007) J Catal 251:249
4. Carroll NJ, Rathod SB, Derbins E, Mendez S, Weitz DA, Petsev DN (2008) Langmuir 24:658
5. Titulaer MK, den Exter MJ, Talsma H, Jansen JBH, Geus JW (1994) J Non-Cryst Solids 170:113
6. Iler RK (1979) The chemistry of silica: solubility, polymerization, colloid and surface properties, and biochemistry, 1st edn. Wiley, New York
7. Lin HP, Mou CY (2002) Acc Chem Res 35:927
8. Yanagishita T, Tomabechi Y, Nishio K, Masuda H (2004) Langmuir 20:554
9. Hosoya K, Bendo M, Tanaka N, Watabe Y, Ikegami T, Minakuchi H, Nakanishi K (2005) Macromol Mater Eng 290:753
10. Kandori K, Kishi K, Ishikawa T (1992) Colloid Surf 62:259
11. Fuchigami T, Toki M, Nakanishi K (2000) J Sol-Gel Sci Techn 19:337
12. Dragosavac MM, Holdich RG, Vladisavljević GT (2011) Ind Eng Chem Res 50:2408
13. Brunauer S, Emmett PH, Teller E (1938) AIChE 60:309
14. Lam KF, Yeung KL, McKay G (2006) Langmuir 22:9632
15. Dragosavac MM, Sovilj MN, Kosvintsev SR, Holdich RG, Vladisavljević GT (2008) J Membr Sci 322:178
16. van der Graaf S, Steegmans MLJ, van der Sman RGM, Schroën CGPH, Boom RM (2005) Colloid Surf A 266:106

Progr Colloid Polym Sci (2012) 139
DOI: 10.1007/978-3-642-28974-3_3

Dynamic Phenomena in Complex (Colloidal) Plasmas

Céline Durniak[1], Dmitry Samsonov[1], Sergey Zhdanov[2], and Gregor Morfill[2]

Abstract Complex plasmas consist of micron sized microparticles immersed into ordinary ion-electron plasmas. They are model systems to study dynamic phenomena at the kinetic level since the damping due to collisions with neutrals is several orders of magnitude smaller than in colloids. Here we report on experimental observations and numerical simulations of dynamic phenomena in complex plasmas: shocks, tsunami effect, and soliton collisions. Shocks, propagating discontinuities, are excited by large amplitude voltage pulses. As they propagate, they melt the lattice, which then re-crystallizes. The propagation of a pulse in an inhomogeneous lattice induces an increase of the amplitude of the pulse ("tsunami effect") even in the presence of damping. The interaction between counterpropagating solitons is shown to be influenced by the lattice structure.

Introduction

Complex or dusty plasmas are ordinary ion-electron plasmas with added microparticles or grains [1]. These microparticles are charged by collisions with electrons and ions, predominantly negatively due to higher mobility of the electrons. They interact with each other electrostatically via a Yukawa potential and often form ordered structures. Similar to colloids, complex plasmas can exist in solid, liquid or gaseous states and exhibit phase transitions. They can be defined as the plasma state of soft matter [2]. In cases of very low thermal energy compared to the potential energy of interaction (strong coupling regimes), crystalline structures are formed, which are often called "plasma crystals". These two-dimensional crystal lattices have a hexagonal structure. Complex plasmas are characterized by the screening parameter $\kappa = a/\lambda_D$, where λ_D is the Debye screening length and a is the average interparticle spacing (usually of the order of $0.1-1$ mm).

They can be found in space [3], *e.g.* in planetary rings [4], comets or interstellar clouds. In plasma technology, dust contamination has negative effects on the yield of semiconductor devices [5]. However beneficial applications include particle growth, which can be used as fine powders or as quantum dots in nanoelectronics or for biomedical applications [6, 7].

Complex plasmas can be obtained in laboratories by adding monodisperse micron-sized grains to a gas discharge, where they can be easily manipulated by changing the gas pressure or the electromagnetic field. Unlike in colloids the grains are weakly damped by gas friction. Therefore dynamic phenomena, such as grain mediated waves [8], shock waves [9, 10], Mach cones [11, 12], and solitons [13, 14], can be observed in complex plasmas at the kinetic level.

Here we present the results of our experiments and molecular dynamics simulations on wave propagation in complex plasmas. The collision of two counter-propagating dissipative solitons was studied as well as the propagation of dissipative solitons in an inhomogeneous lattice and of high amplitude perturbations that are shock waves.

Experimental Setup

The experiments were performed in a capacitively coupled radio-frequency (rf) discharge chamber as shown in Fig. 1a, b. Argon flow of either 0.5 sccm for the shock experiment or 1.2 sccm for the tsunami experiment maintained the working gas pressure in the chamber of 1.8 or 0.682 Pa respectively. An rf power of either 10 W (shock) or 2 W (tsunami) was applied to the lower disc electrode which was 20 cm in

C. Durniak (✉)
[1]Department of Electrical Engineering and Electronics, The University of Liverpool, Brownlow Hill, Liverpool L69 3GJ, UK
e-mail: celine.durniak@liv.ac.uk
[2]Max Planck Institut für extraterrestrishe Physik, Giessenbachstrasse, Garching 85740, Germany

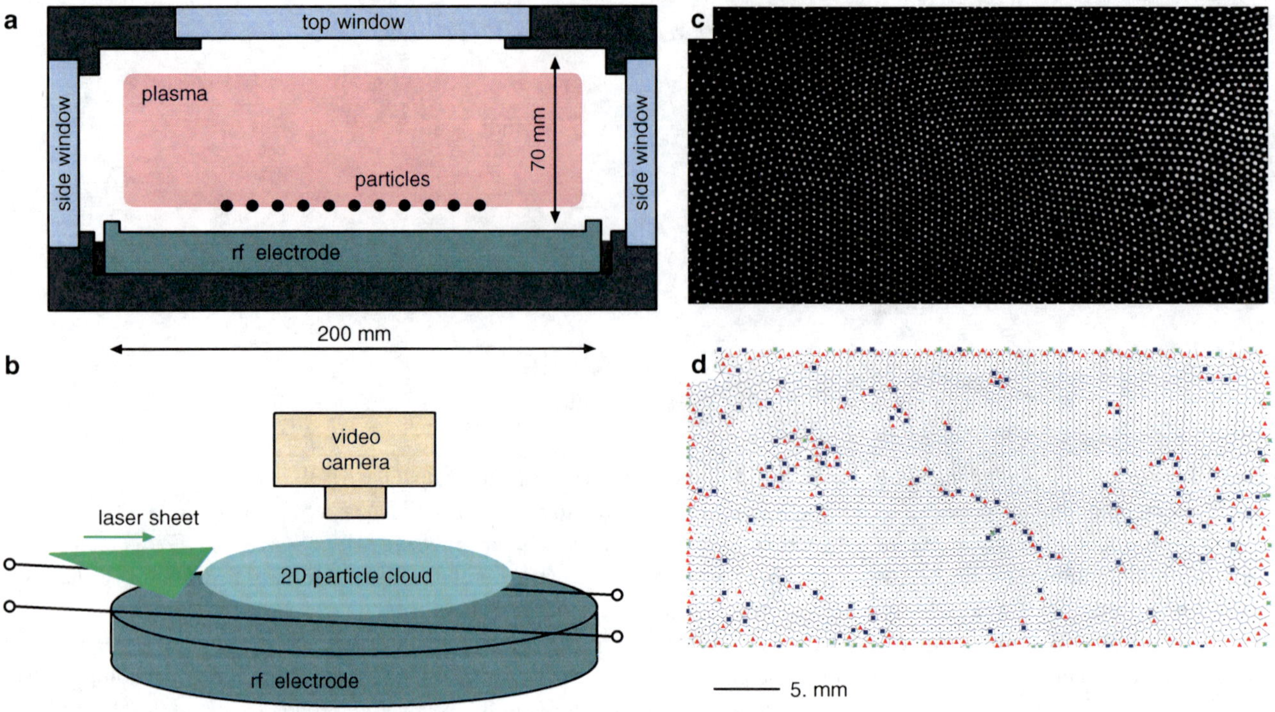

Fig. 1 Experimental setup. (**a**) Side view. Spherical particles charge negatively and form a monolayer levitating in the plasma sheath above the lower electrode. (**b**) Oblique view. Two wires placed below the lattice are used to apply voltage pulses. (**c**) Image of the initial structure of a monolayer hexagonal lattice imaged by the digital video camera (no pulse applied). (**d**) Voronoi map of the hexagonal lattice. The crystal defects are marked with *triangles* (fivefold), *squares* (sevenfold), and *asterisks* (other defects)

diameter. The chamber itself was the other grounded electrode. A DC self-bias voltage helped to suspend the particles in the plasma sheath against the gravity.

The particles that were injected into the plasma through a particle dispenser were monodisperse plastic microspheres of 9.19 ± 0.1 μm in diameter with a mass m of 6.1×10^{-13} kg for the tsunami experiment and 8.9 ± 0.1 μm, 5.5×10^{-13} kg for the shock wave. They levitated in the sheath above the lower electrode. They were confined radially in a bowl shaped potential formed by a rim on the outer edge of the electrode generating a monolayer hexagonal lattice of approximately 6 cm in diameter (Fig. 1c). The particles were illuminated by a horizontal thin sheet of laser light and imaged by a top-view digital camera at a rate of either 102.56 (shock) or 1,000 (tsunami) frames per second.

Two parallel horizontal tungsten wires, both 0.1 mm in diameter were placed below the particle layer at equidistance from the middle of the lower electrode. Short negative pulses with amplitudes of respectively -100 V (shock) and -22.5 V (tsunami) were applied to a wire to excite a single compressional disturbance.

In order to analyze the experimental data, the positions of all particles were identified and they were traced in consecutive frames to calculate the particle velocities.

Molecular Dynamics Simulation

Molecular Dynamics (MD) simulations have been used to study different phenomena in colloid-like systems: glass transitions in colloidal suspensions [15], the driven phases of particles [16], dynamic properties of a fluid confined in a disordered porous system [17], the aggregate structure of surfactants [18] or the relationships between structural aspects of colloid fluids under shear and the shear rate-dependent viscosity [19].

Using a MD code described in [20], we simulated a monolayer complex plasma consisting of 3,000 negatively charged microparticles interacting via a Yukawa potential [21]. The code solved the equations of motion for each microparticle taking into account Yukawa-interaction potential, global parabolic confinement and neutral gas drag. It did not include any explicit plasma, which was taken into account only as the confining and interaction potentials. The grain charge and the screening or Debye length λ_D were kept constant throughout the simulations. The grains were horizontally confined by a parabolic confining potential.

We ran our simulations with the following parameters: a grain mass $m = 5 \times 10^{-13}$ kg, a damping coefficient of 1 Hz, a Debye length $\lambda_D = 1$ mm, and a grain charge Q of $16{,}000e$.

The particles were first placed randomly into the simulation box. Each particle then interacted with all the other particles and the code was run until the equilibrium was reached and a monolayer crystal lattice was formed. The lattice was then excited by a pulsed force applied inwards at one or both sides of the lattice at a fraction of the lattice diameter from the center. The temporal shape was a truncated parabola with a duration of 130 ms (defined as the full width at half maximum amplitude of the pulse). The spatial profile in the x direction was either Gaussian or half-infinite with a Gaussian transition. It did not depend on the y coordinate (parallel to the pulse front). The amplitude of the excitation force was expressed in terms of the parameter $4\pi\varepsilon_0 m\lambda_D^2/Q^2$. The main difference with simulating colloids is that by taking into account the influence of the background, *i.e.* the fluid solvent for colloids, the inertial term can be neglected as the system becomes overdamped [15].

In order to analyze the numerical data, the particle positions and velocities were recorded during the simulations. The local two dimensional number density was determined as the inverse area of the Voronoi cells (see Fig. 1c for an example generated in the experiment).

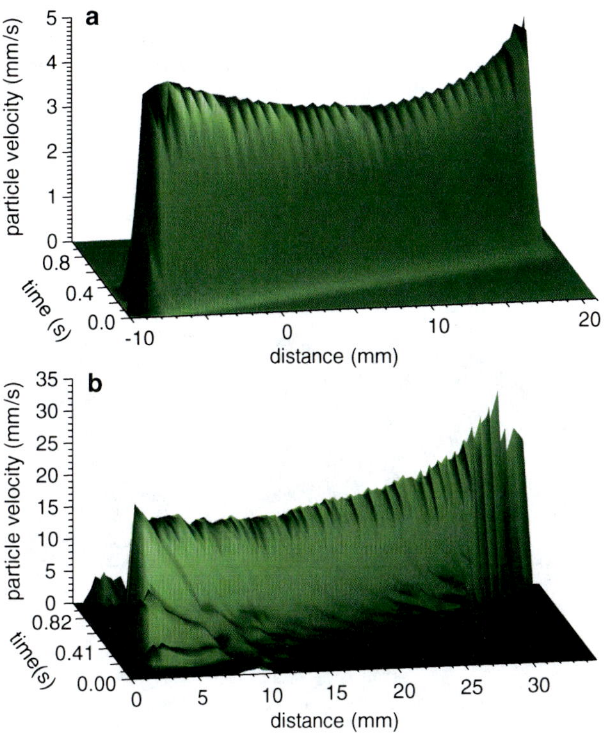

Fig. 2 Tsunami in a complex plasma monolayer. Three dimensional visualization of solitons (particle flow velocity v_x versus propagation distance and time) propagating in an in-homogeneous lattice (**a**) in the simulation and (**b**) in the experiment. The force was applied at $x=-12$ mm in the simulation and $x=0$ mm in the experiment

Dynamic Phenomena

"Tsunami" Effect

The amplification of the pulse amplitude was studied as it reached the inhomogeneous edge of the lattice, *i.e.* with a decreasing number density, even in the presence of damping.

The experimental chamber was slightly tilted in order to maximize the lattice inhomogeneity on one side. Single pulses were excited in both the experiment and the simulation by either applying a voltage pulse to only one of the wires or an excitation force to one of the sides of the lattice. The spatial profile of the simulated excitation force was half-infinite in the x direction with a Gaussian transition [*i.e.* exp $(-(x-x_0)^2/w^2)$ for $x \geq x_0$ and 1 for $x < x_0$] with $x_0=-12$ mm and a waist w of 2 mm. The simulation was performed with a force amplitude $F_{ex0}=1$ a.u.

Figure 2a shows the particle velocity v_x in the x direction. The crystal had a screening parameter $\kappa=0.575$ and a diameter of 41.5 mm. In this crystal, the number density varied between 3.71 mm^{-2} at the center of the crystal to 1.55 mm^{-2} at the edge ($x_e=17.4$ mm). The absolute value of the density gradient increased from 0.07 mm^{-2}/mm at $x=5$ mm to 0.25 mm^{-2}/mm at x_e. The maximum of the particle velocity first decreased due to damping but then it increased as the soliton reached the edge of the lattice where the number density was smaller. The amplitude of the pulse increased by a factor of 1.14 between $x=0$ and x_e. We found that the pulse propagation velocity C varied between 51.34 mm/s in the middle of the crystal ($x=0$) to 35.46 mm/s at the edge. Comparing this velocity to the dust lattice wave speed C_{DL}, the soliton Mach number $M=C/C_{DL}$ decreased from 1.28 to 1.25. As the amplitude of the pulse increased towards the edge of the lattice, we observed a steepening of the front pulse by 37% as well as a decrease of its width by 26%.

The same effect has been found in the experiment, which was performed with a 5 s, −22.5 V excitation pulse as shown in Fig. 2b. In [22] we also reported that this amplification was in qualitative agreement with Korteweg de Vries soliton solution.

Head-On Collision of Counterpropagating Solitons

Collisions between same amplitude solitons were studied in [23] and revealed that the solitons experienced a temporal delay after their interactions, which increased with the amplitude of the excitation force. We also observed that the

Table 1 Simulated collision – characterization of the interaction of two counterpropagating pulses. The amplitude, calculated at the collision point, corresponds to the maximum number density n. We also report the time the collision occurred, its position X in the direction perpendicular to the wires ($X=0$ corresponds to the middle of the lattice), and the pulse velocity C before the collision for $0.4 \leq t \leq 0.65$ s. $F_{ex0,\mathrm{left}} - F_{ex0,\mathrm{right}}$ corresponds respectively to the amplitude of the excitation pulse applied on the left or right hand side of the lattice

Numerical run	Excitation force (a.u.)		Amplitude n (mm/mm^2)	Interaction time T (s)	Interaction position X (mm)	Pulse velocity (mm/s)	
	$F_{ex0,\ \mathrm{left}}$	$F_{ex0,\ \mathrm{right}}$				C_{left}	C_{right}
A	2	2	1.49	0.795	0.09	23.8	−23.7
B	2	3	1.91	0.735	−1.14	23.8	−27.3
C	3	2	1.76	0.745	1.14	26.9	−23.7

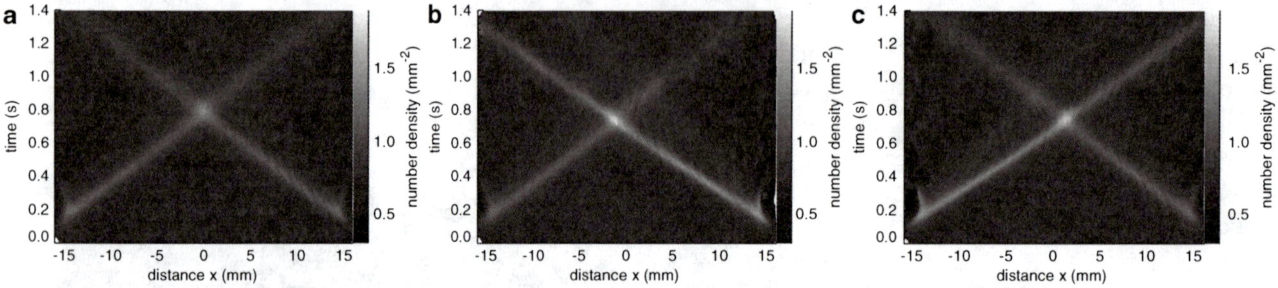

Fig. 3 Simulated collision – trajectory of interacting waves produced from the number density n. The excitation amplitudes were (**a**) the identical and equal to $F_{ex0}=2$ a.u. (run A of Table 1), (**b**) different with $F_{ex0,\mathrm{left}}=2$ and $F_{ex0,\mathrm{right}}=3$ a.u. (run B), (**c**) $F_{ex0,\mathrm{left}}=3$ and $F_{ex0,\mathrm{right}}=2$ a.u. (run C)

amplitude of the resulting pulse at the collision point was different from the sum of the two pulse amplitudes.

In this section we numerically investigate the influence of using different initial excitation amplitudes. The two dimensional simulation was performed with a confining parameter of 0.5 Hz. The excitation forces had identical Gaussian profiles along the x axis with a waist of 2 mm and different initial amplitudes as specified in Table 1: (A) a symmetrical case with $F_{ex0}=2$ a.u. and two asymmetrical cases with (B) $F_{ex0}=2$ a.u. and $F_{ex0}=3$ a.u. for respectively the left (propagating towards $x>0$) and right (propagating towards $x<0$) pulses and (C) the mirrored case. They were applied at ± 17 mm from the center of the crystal, which was 93 mm in diameter and had a screening parameter $\kappa=1.325$.

Figure 3 shows grayscale maps of the particle number density n as a function of the distance perpendicular to the wires x and time for the different initial configurations. The pulses propagated from opposite sides of the lattice inward; their amplitudes decreased because of the neutral damping. Their evolution led to a collision in the middle of the lattice for the symmetrical case (Fig. 3a). As the higher amplitude pulse propagated 13–15% faster than the smaller amplitude pulse before the collision, the interaction occurred slightly off center for the asymmetrical cases (Fig. 3b, c and Table 1) and the amplitude of the resulting pulse was different by less than 9%. These differences between runs B and C were due to the inhomogeneous distribution of defects in the lattice (Fig. 4a). The propagation of the higher amplitude pulse also generated a line of defects corresponding to the compression front, which increased the kinetic temperature (Fig. 4b) in comparison to run A.

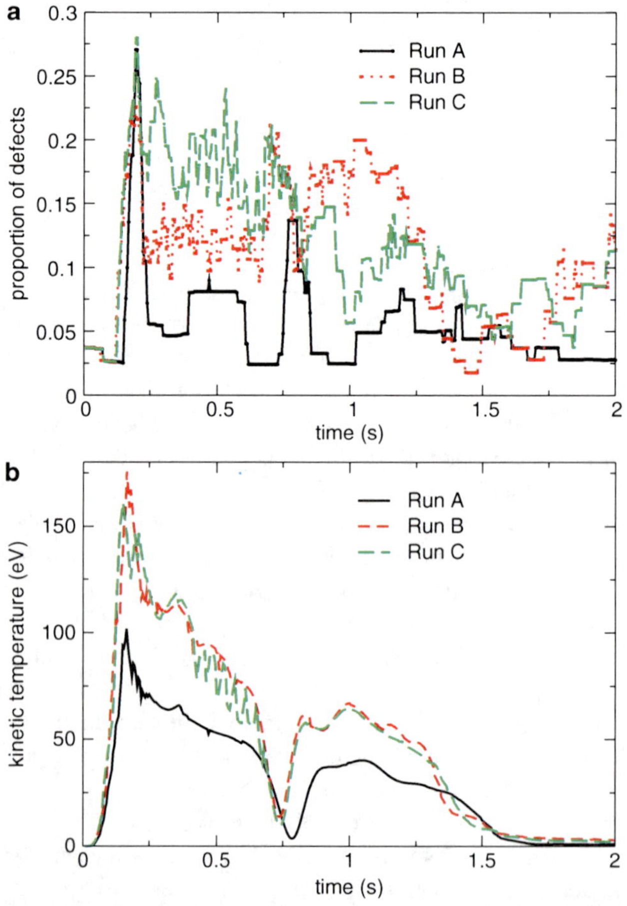

Fig. 4 Simulated collision – temporal evolution of (**a**) the proportion of defects and (**b**) the kinetic temperature for $|x| \leq 16$ mm and $|y| \leq 2.5$ mm. The defects are defined as particles with a number of neighbors different from 6 (hexagonal symmetry)

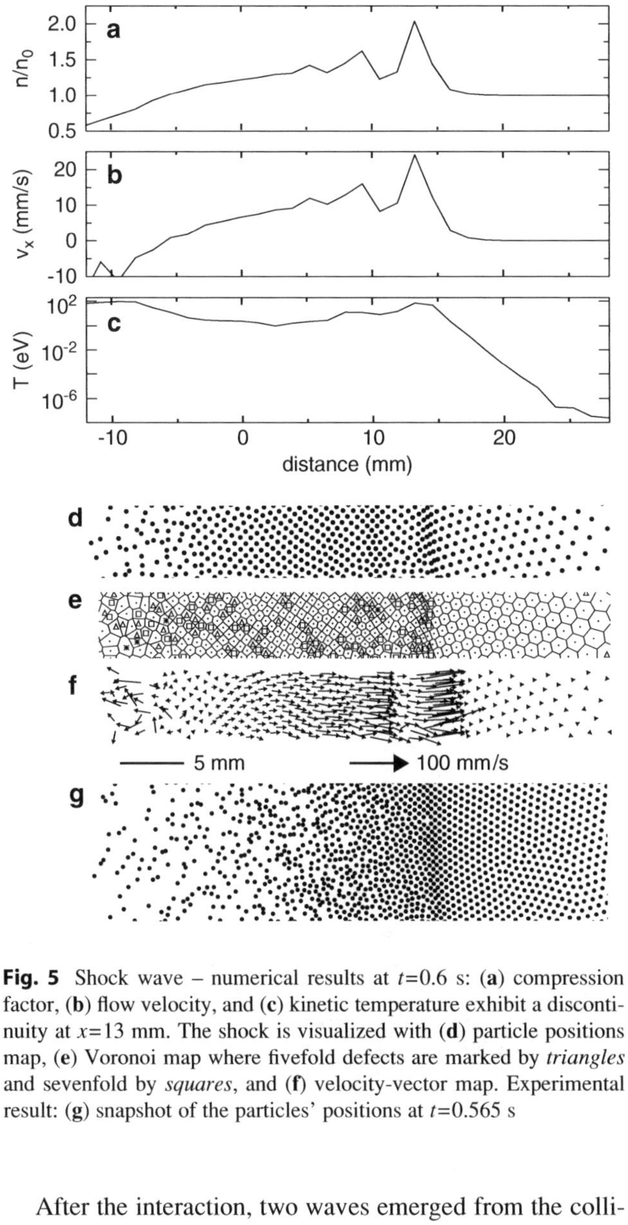

Fig. 5 Shock wave – numerical results at t=0.6 s: (**a**) compression factor, (**b**) flow velocity, and (**c**) kinetic temperature exhibit a discontinuity at x=13 mm. The shock is visualized with (**d**) particle positions map, (**e**) Voronoi map where fivefold defects are marked by *triangles* and sevenfold by *squares*, and (**f**) velocity-vector map. Experimental result: (**g**) snapshot of the particles' positions at t=0.565 s

After the interaction, two waves emerged from the collision point. They were wider and had a reduced amplitude compared to the incident waves.

Shock Waves

Shock waves are propagating disturbances which arise from large amplitude perturbations. Therefore they cannot be treated as linear small amplitude waves [9, 24]. The two dimensional simulation was performed with a confining parameter of 1.5 Hz. The crystal had a screening parameter of 0.825 and a diameter of 57.3 mm. The characteristics of the excitation force were the same as for the tsunami effect (section '"Tsunami" Effect') but with an initial amplitude F_{ex0}=4 a.u.

The structure of the shock is shown in Fig. 5. The shock was characterized by an abrupt change in compression factor n/n_0 (ratio of the wave number density to the unperturbed number density) and flow velocity v_x and an increase in kinetic temperature. The shock propagated from left to right at a velocity between 38 and 51 mm/s, melting the lattice; this resulted in a high number of lattice defects shown on the Voronoi map and velocity randomization. The shock front, which compressed the lattice, had a thickness of a few interparticle distances in the simulation (Fig. 5d) and the experiment (Fig. 5g). The oscillatory structure of the shock ($x \leq$ 13 mm in Fig. 5a, b) was similar to that of [25].

Conclusion

We investigated different examples of dynamic phenomena observed in complex plasmas. The numerical results obtained with our Molecular Dynamic simulation code have been found to be in good agreement with the experiments. First the amplitude of a pulse propagating in a lattice with decreasing number density was amplified towards the edge of the crystal, even in the presence of damping. Then the interaction of counterpropagating pulses with different amplitudes was influenced by the inhomogeneous distribution of defects in the crystal. Finally we reported the propagation of shock waves, which are high amplitude perturbations.

References

1. Morfill GE, Ivlev AV (2009) Rev Mod Phys 81:1353
2. Chaudhuri M, Ivlev AV, Khrapak SA, Thomas HM, Morfill GE (2010) Soft Matter 7:1287
3. Goertz CK (1989) Rev Geophys 27(2):271
4. Havnes O, Aslaksen T, Hartquist TW, Li F, Melandso F, Morfill GE, Nitter T (1995) J Geophys Res 100(A2):1731
5. Selwyn GS, Heidenreich JE, Haller KL (1990) Appl Phys Lett 57 (18):1876
6. Samsonov D, Goree J (1999) J Vac Sci Technol 17(5):2835
7. Boufendi L, Jouanny MC, Kovacevic E, Berndt J, Mikikian M (2011) J Appl Phys D: Appl Phys 44:174035
8. Piel A, Homan A, Klindworth M, Melzer A, Zafiu C, Nozenko V, Goree J (2003) J Phys B: At Mol Opt Phys 36:533
9. Samsonov D, Morfill GE (2008) IEEE Trans Plasma Sci 36 (4):1020
10. Fortov VE, Petrov OF, Molotkov VI, Poustylnik MY, Torchinsky VM, Naumkin VN, Khrapak AG (2005) Phys Rev E 71(3):036413
11. Melzer A, Nunomura S, Samsonov D, Ma ZW, Goree J (2000) Phys Rev E 62(3):4162
12. Samsonov D, Goree J, Ma ZW, Bhattacharjee A, Thomas HM, Morfill GE (1999) Phys Rev Lett 83(18):3649
13. Sheridan TE, Nosenko V, Goree J (2008) Phys Plasmas 15 (7):073703

14. Nosenko V, Goree J, Ma ZW, Dubin DHE, Piel A (2003) Phys Rev E 68(5):056409
15. Löwen H, Hansen J-P, Roux J-N (1991) Phys Rev A 44(2):1169
16. Reichhardt C, Reichhardt CJO (2011) Phys Rev Lett 106 (6):060603
17. Kurzidim J, Coslovich D, Kahl G (2010) Phys Rev E 82(4):041505
18. Burov SV, Obrezkov NP, Vanin AA, Piotrovskaya EM (2008) Colloid J 70(1):1
19. Krekelberg WP, Truskett TM, Ganesan V (2009) Chem Eng Commun 197(1):63
20. Durniak C, Samsonov D, Oxtoby NP, Ralph JF, Zhdanov S (2010) IEEE Trans Plasma Sci 38(9):2412
21. Konopka U, Morfill GE, Ratke L (2000) Phys Rev Lett 84(5):891
22. Durniak C, Samsonov D, Zhdanov S, Morfill G (2009) Eur-Phys Lett 88:45001
23. Harvey P, Durniak C, Samsonov D, Morfill G (2010) Phys Rev E 81(5):057401
24. Samsonov D, Zhdanov SK, Quinn RA, Popel SI, Morfill GE (2004) Phys Rev Lett 92(25):255004
25. Karpman VI (1979) Phys Lett A 71(2–3):163

Progr Colloid Polym Sci (2012) 139
DOI: 10.1007/978-3-642-28974-3_4

Pretreatment of Used Cooking Oil for the Preparation of Biodiesel Using Heterogeneous Catalysis

Kathleen F. Haigh[1], Sumaiya Zainal Abidin[1,2], Basu Saha[3], and Goran T. Vladisavljević[1,4]

Abstract Used cooking oil (UCO) offers a number of benefits for the production of biodiesel because it is a waste material and relatively cheap; however UCOs contain free fatty acids (FFAs) which need to be removed. Esterification can be used to convert the FFAs to biodiesel, and this work has compared two types of heterogeneous catalyst for esterification. An immobilized enzyme, Novozyme 435, was investigated because it has been shown to give a high conversion of FFAs and it has been compared to an ion-exchange resin, Purolite D5081, which was developed for the esterification of UCO for the production of biodiesel. It was found that a conversion of 94% was achieved using Purolite D5081 compared to 90% conversion with Novozyme 435.

Introduction

Biodiesel consists of mono-alkyl esters produced from renewable sources such as vegetable oil or animal fats. The most common process for making biodiesel is the transesterification of vegetable oils with methanol, in the presence of alkaline catalysts to form fatty acid methyl esters (FAME) which is the biodiesel product. A schematic representation of the reaction is shown in Fig. 1. Vegetable oil is an expensive raw material and as a result alternatives have been investigated and these include non-edible oils such as Jatropha Curacas, by-products from oil refining such as palm fatty acid distillate, animal fats, algal oil and used cooking oil (UCO) [1–3].

G.T. Vladisavljević (✉)
[1]Department of Chemical Engineering, Loughborough University, Leics, LE11 3TU, UK
e-mail: G.Vladisavljevic@lboro.ac.uk
[2]Universiti Malaysia Pahang, Lebuhraya Tun Razak, 26300 Gambang, Kuantan, Pahang Darul Makmur, Malaysia
[3]Department of Applied Sciences, London South Bank University, London, SE1 0AA, United Kingdom
[4]Vinca Institute of Nuclear Sciences, University of Belgrade, 522, Belgrade, Serbia

UCO contains free fatty acids (FFAs), which form during cooking [4] and these need to be removed prior to transesterification. Esterification can be used to convert the FFAs to biodiesel [1] and a schematic of this reaction is shown in Fig. 2. Currently most esterification processes use homogeneous catalysts such as sulfuric or sulfonic acid however homogenous catalysts are difficult to separate from the reaction mixture, generate large amounts of waste water, and require expensive materials to prevent associated corrosion [5]. As a result solid acid catalysts such as ion-exchange resins have been investigated with high FFA conversions reported [4, 6].

Enzyme catalysts often result in higher reaction rates at more benign operating conditions when compared with their chemical counterparts. Novozyme 435, Candida antarctica Lipase B immobilized on acrylic resin and has been reported as an effective esterification catalyst including the esterification of FFAs to FAME [2,7]. The aim of this work is to determine if Novozyme 435 can be used to pretreat UCO which contains approximately 6 wt% FFAs and how the catalyst compares with an effective ion-exchange resin. The ion-exchange resin, Purolite D5081 has been developed for the pretreatment of UCO and has been reported to give a high conversion of FFAs in UCO [6].

Experimental

Purolite D5081 was donated by Purolite International and Novozyme 435 was donated by Novozymes UK Ltd. The UCO was donated by GreenFuel Oil Co Ltd., UK and has an FFA content of approximately 6.4 wt %. The fatty acid composition is linoleic acid (C18:2), 43%, oleic acid (C18:1), 36%, palmitic acid (C16:0), 13%, stearic acid (C18:0), 3.8% and linolenic acid (C18:3), 3.6% [6]. Solvents were purchased from Fisher Scientific UK Ltd. The esterification reactions were carried out using a jacketed batch

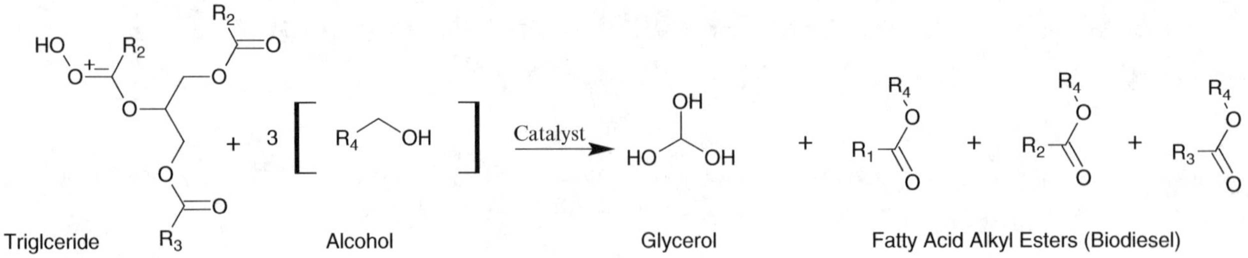

Fig. 1 Schematic representation of the esterification reaction

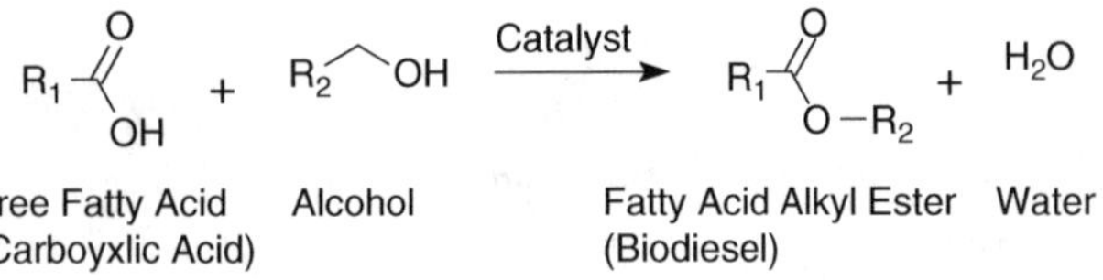

Fig. 2 Schematic representation of the transesterification reaction

Table 1 Summary of the catalyst properties

	Immobilized enzyme Novozyme 435	Ion-exchange resin Purolite D5081
Nature of catalyst	*Candida antarctica* lipase B (CALB) immobilised on acrylic resin	Sulphonated polystyrene cross-linked with divynlbenzene
Physical appearance	White spherical beads	Black spherical beads
Particle size distribution[a]		
d_{10} (μm)	252	396
d_{50} (μm)	472	497
d_{90} (μm)	687	639
BET surface area (m^2/g)	81.6	387
Total pore volume (cm^3/g)	0.45	0.39
Average pore diameter (nm)	17.7	4.1
Porosity (–)	0.349	0.338

[a] d_{x0} is the diameter corresponding to x0 volume % on a relative cumulative particle diameter distribution curve

reactor with a reflux condenser. All samples were analyzed for the FFA content by titration using the ASTM D974 method.

Surface area, pore volume and average pore diameter were determined from adsorption isotherms using a Micromeritics ASAP 2020 surface analyser. In order to investigate internal mass transfer limitations a portion of catalyst was separated into size fractions using a series of sieves on a Fritsch analysette shaker. The amplitude was set to 10 and the catalyst sieved for 120 min. Particle size distribution (PSD) analysis of the original catalyst and sieved fractions were carried out using a Coulter LS 130 Particle Analyzer with isopropyl alcohol as the solvent.

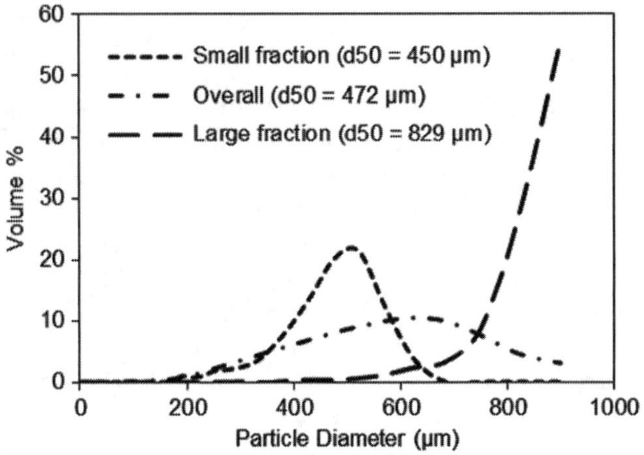

Fig. 3 A comparison of the particle size distributions of Novozyme 435 used to investigate internal mass transfer limitations

Catalyst Characterization Data

A summary of catalyst properties is given in Table 1 and from this data it can be seen that in terms of the particle size distribution Novozyme 435 has a larger spread of particle sizes with the average particle size smaller than Purolite D5081. While Purolite D5081 has a larger surface area than Novoyzme 435 and could have a greater number of accessible catalytic sites, Novozyme 435 is more porous and has a greater pore diameter which may aid the conversion of larger molecules.

Investigation of Internal Mass Transfer Limitations on Conversion

Internal mass transfer resistance is due to the resistance of flow inside the particles and reducing the particle size reduces the diffusion path length and thus internal mass transfer resistance. The internal mass transfer resistance from Novozyme 435 was investigated by sieving the beads into size fractions as shown in Fig. 3. The overall fraction

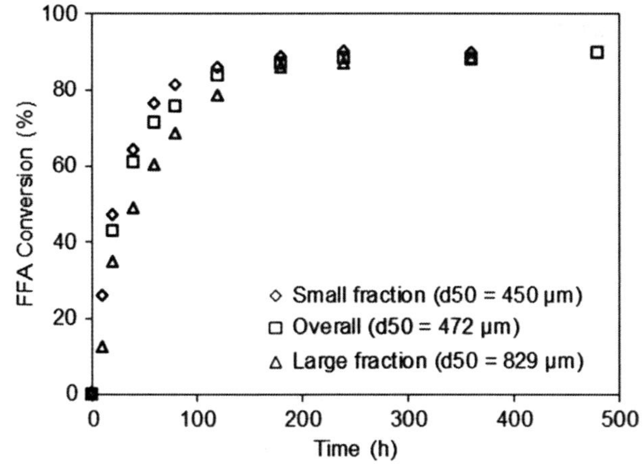

Fig. 4 The effect of particle size distribution of Novozyme 435 on conversion. The reaction conditions are temperature=50°C, catalyst loading=1 wt%, mole ratio=6.2:1 methanol to FFA

(d_{50}=472 μm) represents the size distribution received from the manufacturer.

Figure 4 shows that there are intra-particle diffusion limitations when using Novozyme 435 in the size range supplied by the manufacturer.

Purolite D5081 was previously reported to have no internal mass transfer limitations for the typical particle size supplied by the manufacturer [6]. A sieved fraction with an average particle size (d_{50}) of 463 μm was compared to the original particle size distribution (d_{50}=497 μm).

Comparison of Optimum Conditions for the Two Catalysts Investigated

The optimum conditions for each catalyst were identified by varying the key parameters; for Purolite D5081 these were temperature (50–65°C), methanol to FFAs mole ratio (mole ratio) (66–197:1) and catalyst loading (0.50–1.50 wt%). While for Novozyme 435 the parameters investigated were temperature (30–60°C), methanol to FFAs mole ratio (3.9–15.5:1) and catalyst loading (0.75–1.50 wt%). A summary of the optimum conditions for both catalysts is given in Table 2.

It was found that there was a large difference in the optimum mole ratios for these two catalysts. In the case of Novozyme 435 it was found that there was a narrow range of mole ratio where a high conversion was possible because Novozyme 435 is inhibited by high concentrations of methanol [2] while reducing the concentration means there is insufficient methanol for the reaction. In comparison a much larger mole ratio was required when using Purolite D5081 and the conversion decreased with decreasing mole ratio. There is an order of magnitude difference in the optimum mole ratios and comparing the catalysts at the same reaction conditions would not be meaningful. As a result conversion has been compared using the optimum conditions for each catalyst and this data is shown in Fig. 5.

Table 2 Comparison of the optimum conditions for Novozyme 435 and Purolite D5081

	Novozyme 435	Purolite D5081
Methanol to FFA mole ratio	6.2:1	98:1
Temperature (°C)	50	60
Catalyst loading (wt%)	1.00	1.25

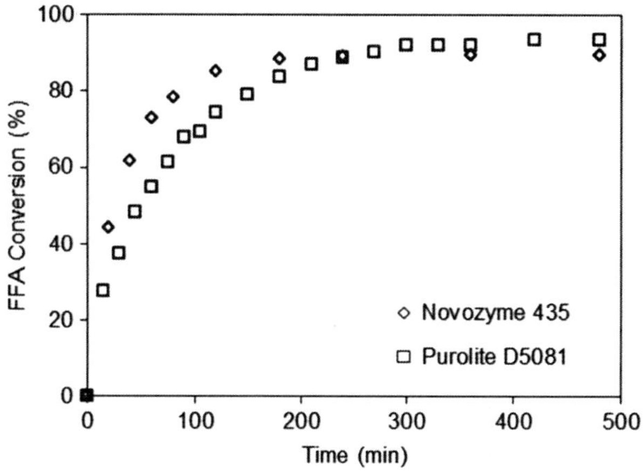

Fig. 5 Comparison of conversion for Novozyme 435 and Purolite D5081 at the optimum conditions for each catalyst as specified in Table 2

From this data it can be seen that with Novozyme 435 the initial reaction rate is faster however the conversion is slightly lower with Novozyme 435 reaching 90% compared to 94% with Purolite D5081 after 600 min of reaction time. Novozyme 435 offers numerous benefits over Purolite D5081 because a high conversion is achieved at a much lower methanol to FFAs mole ratio, lower temperature and catalyst loading. In particular the significant reduction in methanol requirements means that for the same equipment size a much greater capacity is possible and the process will be much safer. A disadvantage is that the cost of enzymes tends to be much greater than that for ion-exchange resins.

Conclusions

The catalytic action of two types of catalyst, an ion-exchange resin, Purolite D5081 and an immobilized enzyme, Novozyme 435 were compared for the esterification pretreatment of UCO for the preparation of biodiesel. Both catalysts gave a good conversion of FFAs to biodiesel in UCO with the optimum conditions summarized in Table 2.

A slightly higher conversion of 94% is possible with Purolite D5081 compared to 90% conversion of FFAs using Novozyme 435 after 600 min of reaction time. It was found that using Novozyme 435 as the catalyst resulted in a large reduction in the amount of methanol required with the optimum mole ratio going from 98:1 methanol to FFA with Purolite D5081 to a mole ratio of 6.2:1.

Acknowledgement We thank EPSRC funding of the PhD scholarship for KH and Universiti Malaysia Pahang and Malaysian Government for the Ph.D. scholarship to SZA. We would also like to thank Purolite International Ltd (Mr. Brian Windsor and the late Dr. Jim Dale) for supplying the ion-exchange catalyst and Novozymes UK. Ltd. (Dr. David Cowan) for supplying the enzyme catalyst.

References

1. Enweremadu CC, Mbarawa MM (2009) Renew Sust Energ Rev 13:2205–2224
2. Talukder MMR, Wu JC, Lau SK, Cui LC, Shimin G, Lim A (2009) Energ Fuel 23:1–4
3. Zabeti M, Wan Daud WMA, Aroua MK (2009) Fuel Process Technol 90:770–777
4. Ozbay N, Oktar N, Tapan N (2008) Fuel 87:1789–1798
5. Caetano CS, Guerreiro L, Fonseca IM, Ramos AM, Vital J, Castanheiro JE (2009) Appl Catal A-Gen 359:41–46
6. Abidin SZ, Haigh KF, Saha B (2012) Ind Eng Chem Res (submitted manuscript)
7. Souza MS, Aguieiras ECG, da Silva MAP, Langone MAP (2009) Appl Biochem Biotech 154:74–88

Progr Colloid Polym Sci (2012) 139
DOI: 10.1007/978-3-642-28974-3_5

Spironolactone-Loaded Liposomes Produced Using a Membrane Contactor Method: An Improvement of the Ethanol Injection Technique

A. Laouini[1,2], C. Jaafar-Maalej[1], S. Gandoura-Sfar[2], C. Charcosset[1], and H. Fessi[1]

Abstract Spironolactone, a hydrophobic drug, has been encapsulated within liposomes using two different preparation methods: the ethanol injection technique and the membrane contactor. The effects of the technique on the prepared liposomes characteristics have been investigated. For this aim, the spironolactone-loaded liposomes were characterized in terms of size, zeta potential, microscopic morphology, encapsulation efficiency and in vitro release profile. Results indicated a significant influence of the applied preparation method. Indeed, when the membrane contactor method was used, the mean size was smaller (123 nm instead of 200 nm), the encapsulation efficiency was higher (93% instead of 80%), and the release profile showed a better dissolution behaviour which may enhance the preparation availability. In conclusion, these results confirmed that the membrane contactor presents an improvement of the ethanol injection technique allowing a continuous production of liposomes at large scale.

Introduction

Over the last decades, advances in pharmaceutical science and technology have facilitated the availability of an extensive range of novel drug carriers including nanoparticles, nanocapsules and liposomes. Due to their biocompatibility, biodegradability and low toxicity, liposomes are extensively studied as drugs and bioactive molecules delivery systems [1, 2].

Since the first report and definition of lipid vesicles by Bangham [3], numerous preparation processes have been developed. Hence, liposomes can be obtained by means of several methods, each of which yields vesicles with special characteristics. However, most techniques are limited in terms of scaling-up ability from the laboratory level to the industrial production.

One of the simplest methods which can be used for liposomes production at large scale is the so-called ethanol injection [4]. Briefly, an ethanolic solution containing the lipid mixture is injected rapidly into an aqueous solution. From the manufacturing point of view, this technique does fulfil the need of a rapid, simple, easily scalable and safe preparation technique. Hence, novel approaches based on the principle of the ethanol injection technique as the microfluidic channel method [5–7] and the cross-flow injection techniques [8] were recently reported.

The membrane contactor based method was already reported for liposome preparation [9]. In this method, the lipidic phase permeated thorough the membrane pores into aqueous phase. Substantial progress was thus achieved, leading from the conventional batch process to a potential large scale continuous procedure.

The present study investigated the preparation of spironolactone-loaded liposomes by ethanol injection technique using syringe and membrane contactor module. The aim of this work was to establish the influence of the preparation technique on liposome characteristics. The selected parameters were the mean size, the zeta potential, the microscopic morphology, the encapsulation efficiency and the in vitro drug release profile.

Materials and Methods

Materials

Reagents

Lipoid E80 was obtained from Lipoïd GmbH (Ludwigshafen, Germany). Spironolactone, cholesterol and phospho-

C. Charcosset (✉)
[1]Laboratoire d'Automatique et de Génie des Procédés (LAGEP), UMR 5007, CNRS, Université Claude Bernard Lyon 1, Bâtiment 308 G, ESCPE Lyon, 2ème étage, 43 bd du 11 Novembre 1918, 69622, Villeurbanne Cedex, France33 (0) 4 72 43 18 6733 (0) 4 72 43 16 99
e-mail: charcosset@lagep.univ-lyon1.fr
[2]Laboratoire de Pharmacie Galénique, Faculté de pharmacie, Rue Avicenne, Monastir 5000, Tunisie

tungstic acid were supplied by Sigma-Aldrich Chemicals (Saint Quentin Fallavier, France). All reagents were acquired with their analysis certificate. Organic solvents (ethanol 95% and chlorhydric acid HCL 37% w/w) were supplied by Carlo Erba Reagenti (Milano, Italy) and were of analytical grade, then used such as without further purification. Ultra-pure water was obtained from Millipore Synergy system (Ultrapure Water System, Millipore).

Syringe Infusion Pump

The syringe infusion pump 22 was purchased from Harvard Apparatus (Holliston, Massachusetts, United States). This precision pump set the industry standard.

Hollow Fiber Module

The Liqui-cel Mini Module X50 was purchased from Alting (Hoerdt, France). This module contains 2,300 polypropylene hollow fibers distributed in a uniform way around a central tube, so as to allow the use of the total membrane surface. The hollow fiber module dimensions are as follow: an inner diameter of 220 μm, an outer diameter of 300 μm, a porosity of 40% and a pore size estimated at 40 nm. The fiber length is 0.115 m and the total active membrane surface is 0.18 m^2.

Methods

Liposome Preparation

Liposome Preparation by the Ethanol Injection Method

Liposomes were prepared by a modified ethanol injection technique [10]. A recent study [11] was carried in our laboratory in order to point out the influence of process and formulation parameters on liposome characteristics. The results showed that the optimum parameters were: aqueous to organic phase volume ratio = 2, stirring rate of aqueous phase during solvent injection = 800 rpm, phospholipid concentration = 20 mg/ml and cholesterol concentration = 20% w/w. The injection velocity has no significant effect on the vesicle mean size. Based on these results, the required amounts of phospholipids, cholesterol and spironolactone (respectively 1 g, 200 mg and 150 mg) were dissolved in 50 ml of ethanol. The resulting organic phase was injected by means of the syringe infusion pump in a defined volume of ultra pure water (100 ml) under magnetic stirring (800 rpm) (RW 20, Ika-Werk). The injection rate was 7 ml/min. Spontaneous liposome formation occurred as soon as the organic solution was in contact with the aqueous phase. This formation became evident on the appearance of the characteristic opalescence of colloidal dispersions. Then, the liposomal suspension was stabilized for 15 min under magnetic stirring. Finally the ethanol was removed by evaporation under reduced pressure (Rotavapor R-144, Buchi, Flawil, Switzerland).

Liposome Preparation by the Membrane Contactor Method

A schematic diagram of the experimental set-up used is shown in Fig. 1.

The system included a positive displacement pump (Filtron, France), a pressurized vessel (equipped with a manometer M_3) connected on one side to a nitrogen bottle (Linde Gas, France) and on the other side to the hollow fiber module (with two manometers M_1 and M_2, respectively placed at the inlet and outlet of the device).

A recent study was conducted by Laouini et al. [12] to determine both process and formulation factors effects on liposome suspension which is obtained by the membrane contactor module. Results showed that the optimum parameters were: aqueous to organic phase volume ratio = 2, organic phase pressure = 1.8 bar, phospholipid concentration = 20 mg/ml and cholesterol concentration = 20% w/w.

For liposome preparation, the required amounts of phospholipids, cholesterol and spironolactone (respectively 5 g, 1 g and 750 mg) were dissolved in 250 ml ethanol. The organic phase was placed in the pressurized vessel. The connecting valve to the nitrogen bottle was opened and the nitrogen pressure was set at a fixed level. The aqueous phase (500 ml) was then pumped through the membrane contactor module using the positive displacement pump. When the water arrived to the inlet of the hollow fiber module, the valve connecting the pressurized vessel to the filtrate side of the membrane device was opened so that the organic phase permeated through the pores of the hollow fibers into the aqueous phase. Spontaneous liposome formation occurred as soon as the organic solution was in contact with the aqueous phase. The experiment was stopped when air bubbles started to appear in the tube connecting the pressurized vessel to the membrane module, indicating that the pressurized vessel was empty. Finally, the liposomal suspension was stabilized and the ethanol removed as previously described.

Liposome Characterization

In order to assess the liposome quality and to obtain quantitative measurements which allow comparison between different liposome batches, measured parameters were:

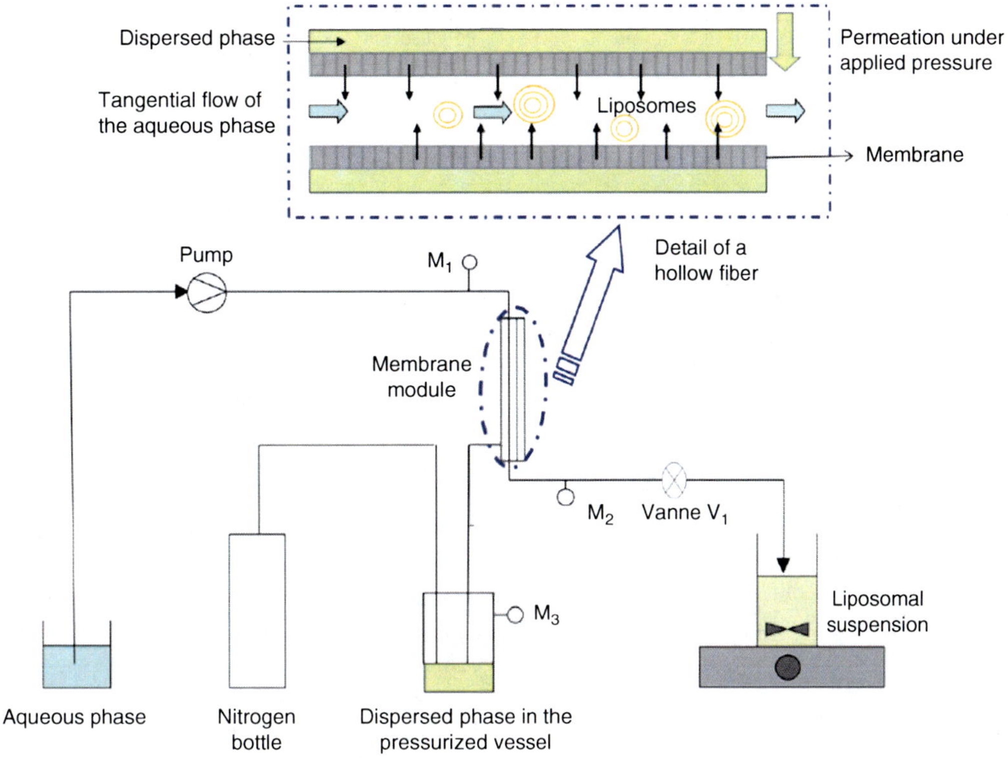

Fig. 1 Schematic diagram of the experimental set-up

Size Analysis

Dynamic light scattering (DLS), otherwise known as photon correlation spectroscopy (PCS), is extensively used in liposome size distribution analysis [13–15]. In this study, a Malvern Zetasizer Nano-series (Malvern Instruments Zen 3,600, Malvern UK) was used. Each sample was diluted 100-fold with ultra-pure water and analyzed in triplicate at 25°C. The data on particle-size distribution were collected using the DTS (nano) software (version 5.0) provided with the instrument.

Zeta Potential Determination

Measurements of zeta potential are commonly used to predict the colloidal system stability. The zeta potential was determined using a Malvern Zetasizer Nano-series (Malvern Instruments Zen 3,600, Malvern UK) and measurements were performed at least three times after dilution in water. The zeta potential was calculated from the electrophoretic mobility applying the Helmholtz-Smoluchowski equation [16].

Microscopic Observation

Transmission electron microscopy (TEM) images were taken using a CM 120 microscope (Philips, Eindhoven, Netherlands) operating at an accelerating voltage of 80 kV. A drop of liposome suspension was placed onto a carbon-coated copper grid; the suspension excess was removed with a filter paper leaving a thin liquid film stretched over the holes. Negative staining using a 2% phosphotungstic acid solution (w/w), pH 7.1, was directly made on the deposit during 1 min. Finally, the excess of phosphotungstic solution was removed with a filter paper and stained samples were observed.

Encapsulation Efficiency

Liposome preparations are a mixture of encapsulated and free drug fractions. Methods for determining the amount of encapsulated material within liposomes typically rely on destruction of the lipid bilayer and subsequent quantification of the released material.

Spironolactone concentrations were measured at λ absorbance of 237 nm with a spectrophotometer UV-vis (Shimadzu UV mini-1,240 V, Kyoto, Japan). The spectrophotometric analytical method was validated as usually required (data not shown).

In the present study, the liposome encapsulation efficiency was determined from the amount of entrapped drugs using the ultracentrifugation technique. Briefly, total spironolactone amount (TSA) was determined after having dissolved and disrupted drug-loaded liposomes in ethanol using

Table 1 Spironolactone-loaded liposomes mean size, polydispersity index and zeta potential

Used preparation method	Mean size ± S.D.[a] (nm)	Polydispersity index ± S.D.[a]	Zeta potential ± S.D[a](mV)
Ethanol injection[b]	200 ± 4	0.310 ± 0.020	−20.5 ± 0.9
Membrane contactor[b]	123 ± 3	0.190 ± 0.030	−23.0 ± 0.6

[a]*S.D.* standard deviation (n=3)
[b]The mean of three batches

an ultrasound bath (Bandelin Sonorex, Schalltec GmbH, Germany) for 10 min. Then, spironolactone-loaded liposome sample was centrifuged (Optima™ Ultracentrifuge, Beckman Coulter, USA) at 50,000 rpm for 50 min at +4°C. The free spironolactone amount (FSA) was determined in the supernatant.

The spironolactone encapsulation efficiency (E.E.) was calculated as follows:

$$\text{E.E.} = \frac{TSA - FSA}{TSA} * 100$$

The encapsulation efficiency was determined in triplicate.

In Vitro Drug Release Study

Spironolactone release was evaluated using the dialysis tube technique. The dialysis membrane Spectra/Por 7 (Spectrum Labs, Breda, Netherlands) was selected according to drug permeability so that no spironolactone adsorption occurred on the membrane (molecular weight cut off of 50 kD). A 4 ml aliquot of liposomal suspension was placed in a hermetically tied dialysis bag, and dropped into 1.5 l of an aqueous receptor medium (chlorhydric acid HCl 0.1 N mentioned in the 31st edition of the American Pharmacopeia). Perfect sink conditions prevailed during the drug release studies and the entire system was kept at 37+/−2°C under continuous magnetic stirring at 70 rpm. The receptor compartment was closed to avoid evaporation of the dissolution medium. Three milliliter samples of the dialysate were taken at various time intervals and assayed for spironolactone concentration by spectrophotometric method. The same volume was replaced with fresh dissolution medium so that the volume of the receptor compartment remained constant. All kinetic experiences were conducted in triplicate and the mean values were taken.

Results and Discussion

Mean Size and Zeta Potential

Table 1 presents the mean size and the zeta potential of the liposomes prepared by ethanol injection using syringe and the membrane contactor module. It can be observed that the vesicle size decreased from 200 to 123 nm when using of the hollow fiber module instead of the syringe.

For pharmaceutical and clinical use, liposomes must fulfil several criteria in terms of size and zeta potential [17]. Liu et al. [18] had showed that liposomes smaller than 70 nm are taken up from the blood stream by liver parenchymal cells, while liposomes larger than 200 nm accumulate in the spleen. An optimum size range of 70–200 nm has thus been identified to give highest blood concentration of liposomes. Therefore, it can be considered that liposomes prepared using the membrane contactor method offers better characteristics than those prepared by the ethanol injection technique.

Since it was previously reported by Lyklema and Fleer [19] and Wiacek and Chibowski [20] that a negative zeta potential higher than 20 mV was considered as an optimal potential for preventing vesicle coalescence, the zeta potential values confirm the stability of our preparations (values comprised between −20 and −23 mV)

Microscopic Observation

As shown in Fig. 2, the morphological investigation using transmission electron microscopy revealed that there was no impact of the preparation method on liposome morphological characteristics. In both cases, prepared liposomes obtained were nanometric sized and quasi-spherical shaped. Moreover, no drug crystals were visible on TEM images, regardless the employed preparation technique. However, vesicle membranes were composed of several phospholipids bilayers resulting in multi-lamellar vesicles (i.e., MLVs) and oligo-lamellar vesicles (i.e., OLVs) respectively for liposomes prepared by the ethanol injection technique and the membrane contactor method.

Encapsulation Efficiency

Spironolactone-loaded liposomes were successfully prepared. The encapsulation efficiency was respectively 80 ± 1.3% and 93 ± 1.1% when liposomes were prepared by the syringe injection technique and the membrane contactor

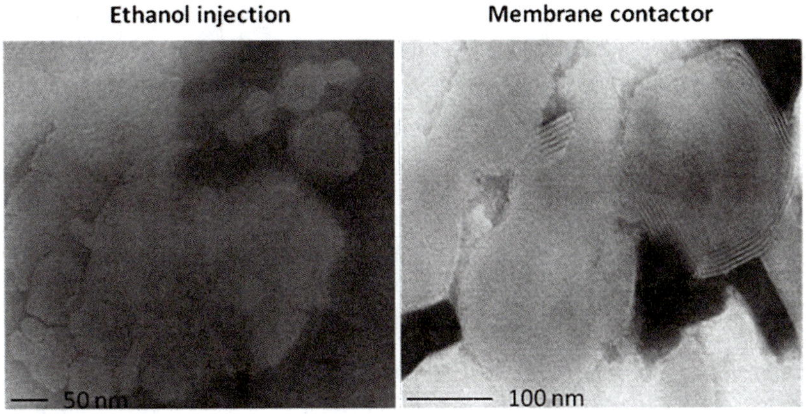

Fig. 2 TEM micrographs of spironolactone-loaded liposomes prepared by the ethanol injection technique and the membrane contactor method

module. The association between the active substance and the carrier (whether the drug is encapsulated in the body or simply adsorbed in the surface) could be assessed by the measurement of the zeta potential [21]. In our study, when using the membrane contactor method, the zeta potential was −43 mV for the drug-free liposomes and −23 mV for the drug-loaded liposomes. The negative surface charge was further shielded in the presence of the drug suggesting that at least a part of the association drug-carrier was surface adsorption and the rest was incorporated within the lipidic matrix. When using the syringe injection technique, no significant change was observed in potential zeta values (respectively −22 mV and −20.5 mV for drug-free and drug loaded liposomes). The liposomes surface properties, remaining unchanged, suggest that spironolactone was mainly entrapped within vesicles bilayers. The higher encapsulation efficiency obtained when the membrane contactor was used, confirms that this method presents a real improvement compared to the classic syringe injection technique. Similar results, not yet published, were recently observed in our laboratory; caffeine encapsulation efficiency within liposomes prepared by syringe injection technique was about 6.9%, whereas membrane contactor module yielded a higher value (11%).

In Limayem-Blouza et al. [22] study, spironolactone loading capacity of nanocapsules was about 90.5% when prepared using a SPG membrane. Several studies [23–25] established that the encapsulation efficiency was proportional to the drug lipophilicity. Thus, the high encapsulation efficiency of spironolactone was believed to be due to its high lipophilicity and therefore its good solubility in the lipid phase.

In Vitro Drug Release Study

As can be seen in the release profile (Fig. 3), when liposomes were prepared using the ethanol injection technique, 54% of spironolactone was released in 60 min. whereas when liposomes were prepared using the membrane contactor method, only 39% of spironolactone was released within 60 min. The release behaviour difference during the first hour could be explained by the important fraction of free drug in the preparation obtained by the ethanol injection technique (20% instead of 7% when liposomes were prepared by the membrane contactor method).

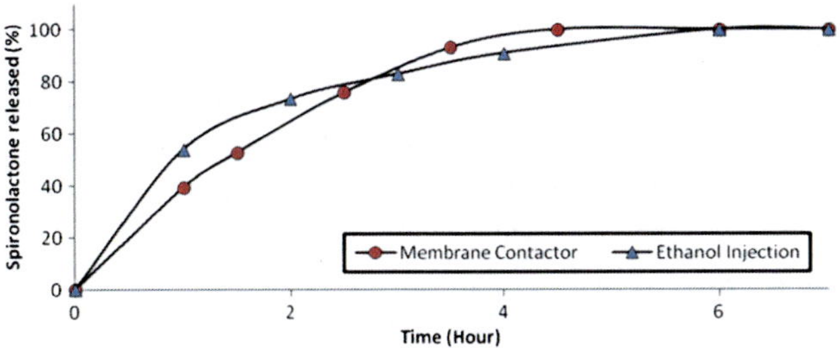

Fig. 3 In vitro spironolactone release profile from liposomes prepared using syringe and membrane contactor module

Otherwise, the dissolution profile showed that spironolactone was completely released after about 6 h from liposomes prepared using the ethanol injection technique and after about 4 h 30 min from liposomes prepared using the membrane contactor method. This rapid release could be ascribed by the difference of liposome mean size, since that in many papers, it has been confirmed that retention time of encapsulated drug within liposomes increased with particles size. According to the Noyes–Whitney equation [26], the drug dissolution rate is directly proportional to its surface area exposed to the dissolution medium. The accelerated release of spironolactone from liposomes prepared by the membrane contactor method (mean size of 123 nm) could be explained by their greater surface area compared to those obtained by the ethanol injection technique (200 nm).

Furthermore, the liposome lamellarity could also explain the release behaviour. Indeed, multi-lamellar liposomes prepared by the ethanol injection technique exhibited reduced release rates compared to oligo-lamellar liposomes obtained by the membrane contactor method.

Conclusion

The purpose of this research was to study the influence of the preparation method on the final characteristics of lipid vesicles. The results showed that liposomes obtained by the membrane contactor method presents better characteristics compared to liposomes obtained by the ethanol injection technique: (1) a smaller mean size (123 nm versus 200 nm) which may lead to a highest blood concentration, (2) a higher encapsulation efficiency (93% versus 80%) and (3) a better release profile and dissolution rate allowing the use of the prepared liposomes for paediatric medication. For the treatment of children heart failure [27], and as it is commercially available only in a solid dosage form and children have difficulty swallowing whole tablets or capsules, spironolactone liquid formulations are preferable [28]. Thus, several extemporaneous formulations have been developed especially suspensions that showed incomplete oral behaviour, slow dissolution rate and a risk of degradation during storage [29]. Spironolactone encapsulation into liposomes may enhance its availability by improving the dissolution rate, and protect the drug from degradation by confining it within lipid vesicles.

On the other hand, the use of the membrane contactor module successfully led to the preparation of 750 ml of liposomes in less than 1 min and 30s, versus a preparation of only 150 ml within 7 min when the syringe injection is used. Therefore, the membrane contactors offer a better efficiency than the classic ethanol injection technique allowing continuous production of liposomes.

In the near future, this study is intended to be completed by a comparative study of the preparations stability.

References

1. Lian T, Ho RJ (2001) Trends and developments in liposome drug delivery systems. J Pharm Sci 90:667–680
2. Torchilin VP (2005) Recent advances with liposomes as pharmaceutical carriers. Drug Discov 4:145–160
3. Bangham AD (1978) Properties and uses of lipid vesicles: an overview. Ann N Y Acad Sci 308:2–7
4. Batzri S, Korn ED (1973) Single bilayer liposomes prepared without sonication. Biochim Biophys Acta 298:1015–1019
5. Jahn A, Vreeland WN, Gaitan M, Locascio LE (2004) Controlled vesicle self-assembly in microfluidic channels with hydrodynamic focusing. J Am Chem Soc 126:2674–2675
6. Pradhan P, Guan J, Lu D, Wang PG, Lee LG, Lee RJ (2008) A facile microfluidic method for production of liposomes. Anticancer Res 28:943–948
7. Vemuri S, Yu C, Wangsatorntanakun V, Venkatram S (1990) Large-scale production of liposomes by microfluidizer. Drug Dev Ind Pharm 16:2243–2256
8. Wagner A, Platzgummer M, Kreismayr G (2006) GMP production of liposomes: a new industrial approach. J Lip Res 16:311–319
9. Jaafar-Maalej C, Charcosset C, Fessi H (2011) A new method for liposome preparation using a membrane contactor. J Lip Res 21:213–220
10. Kremer JMH, Vander Esker MW, Pathmamanoharan C, Wiessema PH (1977) Vesicles of variable diameter prepared by a modified injection method. Biochemistry 16:3932–3935
11. Jaafar-Maalej C, Diab R, Andrieu V, Elaissari A, Fessi H (2010) Ethanol injection method for hydrophilic and lipophilic drug-loaded liposome preparation. J Lip Res 20:228–243
12. Laouini A, Jaafar-Maalej C, Sfar S, Charcosset C, Fessi H (2011) Liposome preparation using a hollow fiber membrane contactor – application to spironolactone encapsulation. Int J Pharm 415:53–61
13. Berger N, Sachse A, Bender J (2001) Filter extrusion of liposomes using different devices: comparison of liposome size, encapsulation efficiency, and process characteristics. Int J Pharm 223:55–68
14. Kölchens S, Ramaswamia V, Birgenheiera J (1993) Quasi-elastic light scattering determination of the size distribution of extruded vesicles. Chem Phys Lipids 65:1–10
15. Provder T (1997) Challenges in particle size distribution measurement past, present and for the 21st century. Prog Org Coat 32:143–153
16. Hunter R, Midmore HZ (2001) Zeta potential of highly charged thin double-layer systems. J Colloid Interf Sci 237:147–149
17. Wagner A, Vorauer-Uhlb K, Katingerb H (2002) Liposomes produced in a pilot scale: production, purification and efficiency aspects. Eur J Pharm Biopharm 54:213–219
18. Liu D, Mori A, Huang L (1992) Role of liposome size and RES blockade in controlling biodistribution and tumor uptake of GM1-containing liposomes. Biochim Biophys Acta 1104:95–101
19. Lyklema J, Fleer GJ (1987) Zeta electrical contributions to the effect of macromolecules on colloid stability. Colloid Surf 25:357–368
20. Wiacek A, Chibowski E (1999) Zeta potential, effective diameter and multimodal size distribution in oil/water emulsion. Colloid Surf A 159:253–261
21. Barratt G (2003) Colloidal drug carriers: achievements and perspectives. Cell Mol Life Sci 60:21–37
22. Limayem-Blouza I, Charcosset C, Sfar S, Fessi H (2006) Preparation and characterization of spironolactone-loaded nanocapsules for paediatric use. Int J Pharm 325:124–131
23. Barenholz Y (2003) Relevancy of drug loading to liposomal formulation therapeutic efficacy. Liposome Res 13:1–8
24. Fresta M, Cavallaro G, Giammona G, Wehrli E, Puglisi G (1996) Preparation and characterization of polyethyl-2-cyanoacrylate nanocapsules containing antiepileptic drugs. Biomaterials 17:751–758
25. Xu Q, Tanaka Y, Czernuszka JT (2007) Encapsulation and release of a hydrophobic drug from hydroxyapatite coated liposomes. Biomaterials 28:2687–2694
26. Mosharraf M, Nystrom C (1995) The effect of particle size and shape on the surface specific dissolution rate of microsized practically insoluble drugs. Int J Pharm 122:35–47
27. Simmonds J, Franklin O, Burch M (2006) Understanding the pathophysiology of paediatric heart failure and its treatment. Curr Paediatr 16:398–405
28. Standing F, Tuleu C (2005) Pediatric formulations: getting to the heart of the problem. Int J Pharm 300:56–66
29. Allen LV, Erickson MA (1996) Stability of ketonazole, metolazone, metronidazole, procainamide, hydrochloride and spironolactone in extemporaneously compounded oral liquids. Am J Health Syst Pharm 53:2073–2078

Progr Colloid Polym Sci (2012) 139
DOI: 10.1007/978-3-642-28974-3_6

The Multiple Emulsion Entrapping Active Agent Produced via One-Step Preparation Method in the Liquid–Liquid Helical Flow for Drug Release Study and Modeling

Agnieszka Markowska-Radomska and Ewa Dluska

Abstract The paper presents a theoretical mass transfer model of the release process of active agent from complex dispersed systems such as multiple emulsions, microemulsions, micro/nanoparticles. The model is characterized by five parameters describing internal structure of the delivery system and the conditions of the release environment which enable the determination of the release rate and its sensitivity to the process parameters. The model was validated by experimental data of an active component release from two sets of multiple emulsions prepared via a novel one-step emulsification method in the continuous Couette-Taylor Flow contactor. The simulation of release profiles confirmed the importance of the internal multiple emulsions structure (drop size, packing volume) as well as the intensity of external mixing in the modeling of the controlled release process. The presented model allowed the mass released to be determined with satisfactory agreement with experimental data after optimization of parameters describing internal emulsions structure.

Introduction

Multiple emulsion systems are more complex type of dispersed system in which the dispersed phase contain smaller droplets of the internal phase that have the same or different composition as the external phase, Fig. 1. The multiple emulsions are considered to be of Oil-in-Water-in-Oil (O/W/O) and Water-in-Oil-in-Water (W/O/W) emulsion systems (Fig. 1a). In O/W/O systems an aqueous phase separates internal and external organic phases. In W/O/W systems organic phase separates internal and external aqueous phases. The phase, which separates two aqueous or organic phases is known as membrane phase and acts as a different barrier and semi-permeable membrane for the drugs or moieties entrapped in the internal phase. The internal and external phases may also have different composition: $O_1/W/O_2$, $W_1/O/W_2$ (Fig. 1b).

In recent years, multiple emulsions have been widely used in numerous fields of applications such as pharmaceutics, cosmetics, food and environmental technologies. The specific intrinsic structure makes them useful e.g. for prolonged or controlled active agent delivery system, separation operations (ELM-emulsion liquid membrane) etc. [1]. Using multiple emulsions we can, e.g. prevent degradation of an active agent, improve dissolutions of insoluble substances, mask the taste and smell of active ingredients and control active agent release into the external phase. In view of the numerous uses of emulsions in different technologies, new methods of their preparation, stabilization, as well as study of mechanisms of active agent (drug) release from multiple emulsion are still under interest of pharmaceutical, medical, chemical and biochemical engineering [1]. Nowadays, in view of the great potential applications of the multiple emulsion products, kinetics of an active agent release is widely studied area. The most common method of multiple emulsions preparation is a two-step stirring emulsification which is widely described in numerous papers [1–3]. The multiple emulsions are prepared also by the phase inversion, the membrane and the microchannel emulsification [4].

Design of multiple emulsion products usually requires expensive experimental investigation. In order to reduce costs, the mathematical models of release process based on the release mechanisms, an intrinsic structure of the product as well as its relationship with the external surrounding acceptor are used. The main drug release mechanisms from multiple emulsions are diffusion of an active agent from internal droplets through the permeable membrane and/or emulsion breakdown/membrane rupture due to shear forces–defragmentation, or as a result of osmotic flow leading to the swelling-breakdown process [1, 5, 6].

A. Markowska-Radomska (✉)
Chemical and Process Engineering, Warsaw University of Technology, Warynskiego 1, Warszawa, 00-645, Poland+48222346325 +48228251440
e-mail: a.markowska@ichip.pw.edu.pl

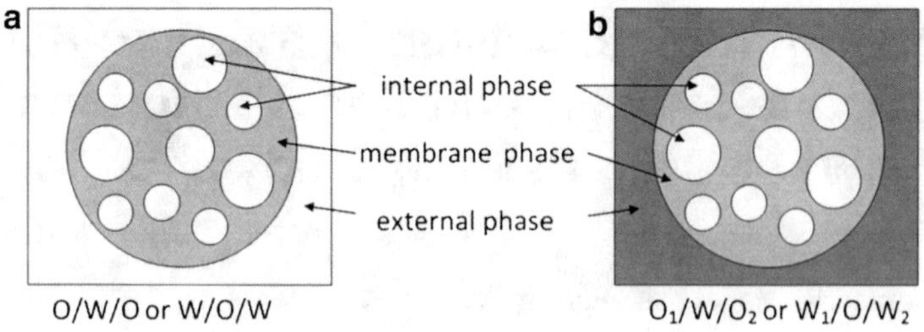

Fig. 1 Types of multiple emulsion: (**a**) the same composition of internal and external phases, (**b**) different composition of internal and external phases

The most commonly reported in the literature models of drug release refer to the release from polymeric microspheres [7–9]. The models for non-eroding porous/macroporous microspheres can be applied to the multiple emulsion for predicting an active agent release kinetics and also to modeled diffusion inside permeable liquid membranes for a lower diffusion coefficient.

The pioneering mass transfer model is the Higuchi model of drug release from solid matrix in contact with a perfect sink where instantaneous drug dissolution and pseudosteady state diffusion are assumed [10, 11]. In the literature, the models considering the Fick's second law with the percolation theory and percolation threshold to estimate the kinetic parameters are reported [12]. Models based on the Fick's second law have been applied for different geometries. These models assumed the geometry of the reservoir system where the drug must diffuse through the layer of a constant thickness or the geometry of the matrix system with drug dissolved or uniformly distributed inside. For matrix system, the radius of the inner interface between the region where the drug is loaded and matrix (diffusing) regions shrinks with time. As the geometry of the release device in the mentioned models is exactly characterized the method of emulsion preparation must be specified. As it was reported in the literature the models considering any geometry are required. This is possible by direct modeling approach which offers cellular automata models for drug release predictions, [13].

In Liao and Lucas paper [14] the mathematical models corresponding to the breakage of the emulsions are discussed. Most analytical models of diffusion controlled release are based on the assumption of steady state diffusion. Moreover, most of models, regardless of the mechanism of release assume infinite mass transfer coefficient in the external environment. In fact the finite mass transfer conditions are often seen in drug release under in vivo experiments. As a result the mass transfer resistance and hence mass transfer coefficient does dependent on the release environment conditions.

In this paper the factors that affect the release rate of active agent (drug) from multiple emulsions including external mass transfer resistance and intrinsic structure of multiple emulsions are demonstrated based on the sensitivity analysis of proposed theoretical mass transfer model for prediction of drug release rate from multiple emulsions [15]. Subsequently the model predictions are compared with experimental data of drug release from $O_1/W/O_2$ emulsions produced by an unconventional one-step emulsification method in the Couette-Taylor flow (CTF) contactor reported in our previous paper [16, 17].

Mass Transfer Mathematical Model

The presented mathematical model of the release process from multiple emulsions corresponds to an active agent (drug) diffusion within the membrane phase drops to the surface as well as diffusion through the boundary layers on the surface of the membrane phase drops and diffusion – convection transport of active agent in the external continuous phase.

Since drug release is controlled by diffusion the effects of coalescence and breakage of the internal droplets are ignored. The physiochemical properties of the emulsion system remain constant. In order to obtain the model solution, knowing the mass transfer coefficients inside and outside of the drops, interfacial area as well as driving force are necessary.

The volumetric mass transfer coefficient of active agent through the membrane phase drops (k_La) can be defined from the Happel sphere in cell conceptual model that enables taking into account effect the internal structure of multiple emulsion:

$$k_La = \kappa = \frac{D}{\delta}a = \frac{3D\phi^{1/3}}{r_i^2(1-\phi^{1/3})} \qquad (1)$$

Where δ (m) is the layer thickness representing mass transfer resistance in the membrane phase, D ($m^2\ s^{-1}$) is the drug molecular diffusivity within the membrane phase drops, r_i (m) is the radius of the internal phase drops, a (m^{-1}) is the interfacial area of internal phase drops.

The release rate modeling was based on mass transfer of an active agent between two continuous phases: the membrane phase and assumed pseudo homogenous phase of the internal droplets. This simplification avoids mathematical calculations of the release from the internal mass sources and assumes that in the membrane phase represented by concentration function of an active agent C(r, t) co-exists the concentration function $C_S(r, t)$ of an active agent in pseudo homogenous internal phase. The driving force was expressed as the difference between the concentration of an active agent at the interface of the internal phase $C_S{}^*(r, t)$ and in the membrane phase C (r, t). The constant concentration in the internal droplets was assumed due their small size and at least one order of magnitude higher value of diffusion coefficient of an active agent compared to that in the membrane phase. This assumption implies that $C_S=C_S{}^*$.

The mass release to the external environment begins after the achievement of an equilibrium state inside the membrane phase drops t: $C_{S,eq}=C_{eq}$ for $t_0=0$. The volume of well-mixed release medium is big enough so that the bulk concentration of a drug is $C_\infty=0$.

With the above assumptions the release process of active agent from multiple emulsions considered as monodisperse population of drops of the internal and membrane phases is described by set of model equations based on the mass balance of active agent in:

– The membrane phase:

$$\frac{\partial C(r,t)}{\partial t} - D\Delta C(r,t) = (k_L a)[C_S^*(r,t) - C(r,t)]\frac{\phi}{1-\phi} \quad (2)$$

where, Δ represents 3D Laplace operator in spherical coordinates reduced due to the symmetry to the term $\frac{1}{r^2}\frac{\partial}{\partial r}\left(r^2\frac{\partial C(r,t)}{\partial r}\right)$;

– Pseudohomogenous internal phase:

$$\frac{\partial C_S^*(r,t)}{\partial t} = (k_L a)[C(r,t) - C_S^*(r,t)] \quad (3)$$

– The boundary condition at the interface between the external and the membrane phases:

$$D\frac{\partial C(r,t)}{\partial r}\bigg|_{r=R} = -h\cdot m\cdot C(r,t)|_{r=R} \quad (4)$$

where, h (ms^{-1}) represents external mass transfer resistance; m ($m=10^3$) is the partition coefficient of an active agent between membrane and external phases; R (m) is the radius of the membrane phase drops,

– The initial condition:

$$C(r,t) = C_S^*(r,t) = C_{S0}^*\cdot\phi \text{ dla} \qquad t=t_0 \quad (5)$$

where, t_0 (s) is the initial release time

For the mathematical simplification the scaling laws for three selected parameters R, C_{S0}, D were used. The system of (2)–(5) are analytically solved to determine the dimensionless mass released at any time M_{t+} from internal droplets to the external phase as a function of the dimensionless concentration profiles in the membrane (C^+) and internal phases ($C_S{}^+$). The mass released to the external environment was calculated as the difference between the dimensionless initially encapsulated mass of the drug in the internal phase $M_{t+}=1$ and the mass remaining in the emulsion droplets.

$$M_{t^+} = 4\pi\left(\frac{\phi}{3} - \int_0^1 [C^+(r^+,t^+)(1-\phi) + C_S^+(r^+,t^+)\phi](r^+)^2 dr^+\right) \quad (6)$$

The results of model simulation are presented as the cumulative fractional mass released outside of the membrane phase drops expressed as a ratio:

$$(M_{t^+}/M_{t^+=1}) = (M_t/M_\infty) = (M_t/M_0) = f(R,r_i,\phi,D,h) \quad (7)$$

Results and Discussion

Preparation of Multiple Emulsion

The target of emulsification in helical flow contactor was stable $O_1/W/O_2$ emulsion consisting of three phases of different composition. Where the O_1 is the internal phase (phenyl salicylate (active agent) dissolved in liquid paraffin), W is the membrane phase (15 wt% gelatin aqueous phase containing 5 wt% sucrose) and O_2 is the external continuous phase (liquid paraffin).

In all of the experiments as model hydrophobic drug was used phenyl salicylate (salol) which is used in veterinary medicine and is being applied as an analgesic, antipyretic and intestinal antiseptic as well as the sunscreen and preservative. The other substances used were gelatin for microbiology uses (pH=4.0–6.0, gel strength 240–270 g Bloom) and dimethyl sulfoxide (DMSO). In this work as a cross-linker, sucrose was used. All chemicals were of analytical grade. The multiple emulsions system based on gelatin was chosen due to possible pharmaceutical applications. It is favorable if biocompatible polymers are used for the production of controlled release devices. Recently the use of gelatin as a polymer in form of liquids, gels and microspheres for controlled release systems has received much attention. Gelatin is relatively cheap and has the potential for use with a variety of medical agents [18].

The multiple emulsions for release study were produced in the Couette-Taylor flow contactor under different operation conditions such as, the gap size, the rotational frequency of

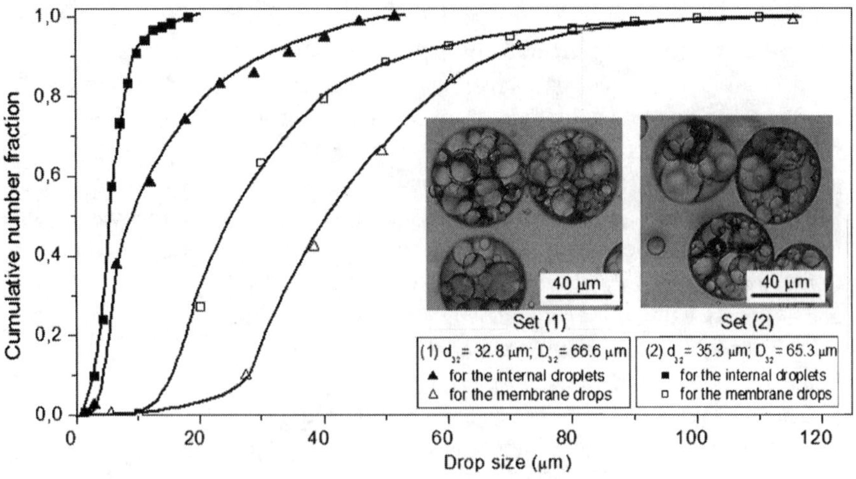

Fig. 2 Cumulative distribution of the number fraction of drops population of $O_1/W/O_2$ emulsions prepared in the CTF contactor

the inner cylinder, the initial concentration of active agent, the volumetric flow rate of oils and water phases. To prepare multiple emulsions, the membrane and internal liquid phases were introduced into the annular gap between the coaxial cylinders of the CTF contactor at the inlet cross section, followed by the external phase. In the CTF contactor run the inner phase was incorporated as the droplets into the membrane phase drops dispersed in the external continuous phase of multiple emulsions. An example of the drop size distribution is illustrated in Fig. 2.

Release Rate Experiments

The two sets of stable $O_1/W/O_2$ emulsions entrapping phenyl salicylate were selected for release experiments, Fig. 2.

The emulsions were characterized by such factors as: the encapsulation efficiency *(EE)*, Sauter mean diameter of the membrane phase drops *(D_{32})* and internal phase droplets *(d_{32})* as well as the ratio of the volume of internal droplets packed in membrane phase drops to the volume of the membrane and internal phase drops (packing volume fraction-ϕ):

- $\text{set}(1): EE = 94.7\%,\ D_{32} = 66.58\mu\text{m},$
  $d_{32} = 32.76\mu\text{m},\ \phi = 0.83,$
- $\text{set}(2): EE = 69.8\%,\ D_{32} = 65.26\mu\text{m},$
  $d_{32} = 35.31\mu\text{m},\ \phi = 0.74,$

Active agent release study was performed in the standard stirred tank (diameter of 0.1 m height of 0.2 m, emulsion volume 0.976 dm^3) at different agitation speeds (100, 250 rpm) and at the constant temperature of 37°C±0.5°C.

To determine the mass of salol released from internal phase of the multiple emulsions to the external paraffin phase, at fixed time intervals the samples of multiple emulsions were withdrawn from the stirred tank and were immediately filtered using hydrophobic membrane to separate the O_1/W drops from the external paraffin phase (O_2). Salol from external paraffin phase was extracted by DMSO (dimethyl sulfoxide) and then analyzed spectrometrically. The mass of released salol was calculated as the mean value from three independent experiments.

For each samples of multiple emulsion withdrawn from the stirred tank during release experiments the drop size distribution was analyzed using the confocal laser scanning microscope LEXT OLS3100 and system optical microscope BX-60 Olympus connected to a digital camera (Olympus). The stability of the multiple emulsions during the release process was controlled on line by the backscattering light monitoring with the Turbiscan Lab and by droplet sizing microscopic observation over time. There were no significantly changes in the drop size distributions and instabilities in the dispersion system (coalescence, osmotic mismatch, change in drop size) during the release process.

All experimental and measuring procedures have been explained in detail in previous works of authors [15–17]. Analysis of the experimental results showed that the release rate profiles are strongly affected by the delivery system structure and related active agent encapsulation, as well as by external environment mixing intensity [15–17].

Comparison of Model and Experimental Data

The experimental active agent release study and modeling for multiple emulsion prepared in helical flow was considered. Two sets of emulsions were chosen for simulations: highly packed emulsions of set 1 and worse emulsions of set 2. The release profiles of salol for $O_1/W/O_2$ emulsions mixed with two different stirring rotations: 100 and 250 rpm are

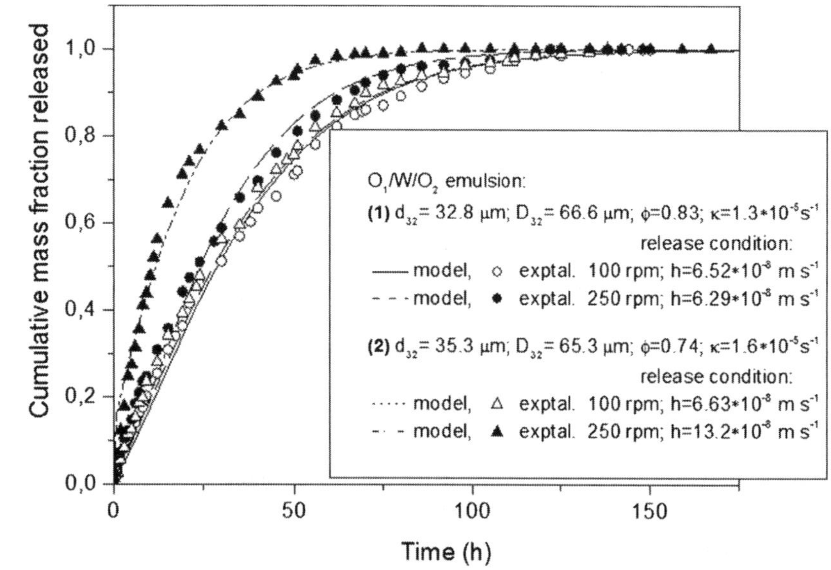

Fig. 3 Simulated and experimental release profiles of salol from multiple emulsion

presented in the Fig. 3. The release profiles show the dependence of the cumulative mass of salol released from O_1/W/O_2 at time *t* to the external continuous phase – (M_t) referred to the mass released at the infinite time (when mass released ≈initial mass encapsulated: $M_\infty \approx M_0$).

The analysis of the model simulations and experimental release profiles was performed for the volumetric mass transfer coefficient ($k_La=\kappa=(D/\delta)a$) expressed by (1) and calculated with other equations such as proposed by Kataoka et al. [19] and Hino et al. [20], where authors considering other ways of expressing the membrane thickness (δ), and Lian et al. [21], where the definition of the mass transfer coefficient was related to the Crank equation of diffusion under the assumption of perfect mixing. As shown analysis in each case to correctly predict the release rate was necessary to optimize the experimental data to determine the volumetric mass transfer coefficient named after optimization parameter (κ). As it turned out the best fitting to the experimental data was obtained when the values of k_La were determined from (1), based on the Happel sphere in cell model.

The parameters κ representing the optimized mass transfer resistance inside membrane phase drops are constant for the specified classes of emulsions and for emulsions with a similar structure and about the same drug loading. As it can be seen in the Fig. 3 the presented model accurately predicts release rate based on the fitted values of the parameter $\kappa(\kappa_{set1} = 1.3 \times 10^{-5}s^{-1}, \kappa_{set2} = 1.6 \times 10^{-5}s^{-1})$ and the values of model parameters (R, r_i, ϕ) taken from release experiments and can be also applied for multiple emulsion prepared by any other methods.

The experimental and prediction release profiles, presented in the Fig. 3, shown that for highly packed structures the external environment mixing not affects release rate and therefore release process is limited by the diffusion resistance within the membrane phase drops. In case of multiple emulsions with less packed structure (below 80%) the influence of the surrounding environment mixing intensity is easily noticeable. d_{32}, ϕ, D, h taken from release experiments accurately predict active agent release kinetics. As the simulations were extended to nano-scale dispersed systems the release profiles predictions with one parameter optimized were also satisfactory. The presented model can accurately predict the release kinetics of a wide range of delivery systems prepared by different methods if structural parameters and active agent loading are including [15].

Conclusion

The theoretical model has been developed to analyze and predict active agent release from multiple emulsions. The presented theoretical model is characterized by five parameters describing emulsions structure and release conditions. For modeling of release rate emulsions structure were characterized by the mean size of membrane and internal phase drops, encapsulation efficiency of active agent and the packing volume. The release conditions included the intensity of external surrounding mixing.

The model was validated by comparison with experiments of active agent (salol) release from emulsions of type O_1/W/O_2 under a variety of conditions. The quantitative analysis of the comparison model simulations of release profiles with corresponding experimental data showed that one of the parameter must be optimized. The parameter estimated from experimental data was the volumetric mass transfer coefficient being in fact parameter of emulsion structure (parameter κ).

The simulations based on the characteristic parameter κ, confirmed the importance of the emulsions structure for the release process modeling. The modeling results strongly suggest that when release experiment is carried out for the highly packed structures (above 80%), the release rate is directly linked to the intrinsic structure of multiple emulsion. For less packed structures the release rate is more sensitive to external environmental mixing intensity.

Knowing the values of optimized parameter κ we can use them for release process designing and computations for multiple emulsion systems with a similar structure and about the same drug loading at any release conditions. This allows the prediction of drug release kinetics without prior in vitro studies. As the simulations were extended to nano-scale dispersed systems the release profiles predictions with one parameter optimized were also satisfactory. The presented model can accurately predict the release kinetics of a wide range of delivery systems prepared by different methods if structural parameters and active agent loading are including.

In addition, the good agreement between experimental data and simulated profiles for structures of different scale (micro and nano), geometry, drug encapsulation and formulation (emulsions, particles) allows the presented model to be considered as a tool to controlled release process design.

This model can be applicable for the systems containing any other active component e.g. flavor dispersion in the gelled emulsions particles, for describing the controlled volatile release process.

Acknowledgments We gratefully acknowledge support from the Polish Ministry of Science and Higher Education – Grant N N209 145836 (2009–2012).

References

1. Aserin A (2008) Multiple emulsions: Technology and Applications Wiley, USA
2. Matsumoto S, Kita Y, Yonezava D (1976) J Colloid Interface Sci 57:353–361
3. Olivieri L, Seiller M, Bromberg L, Ron E, Couvreur P, Grossiord JL (2001) Pharm Res 18:689–693
4. Nisisako T (2008) Chem Eng Technol 31:1091–1098
5. Florence AT, Whitehill D (1981) J Colloid Interface Sci 79:243–256
6. Vladisavljević GT, Shimizu M, Nakashima T (2006) J Membr Sci 284:373–383
7. Arifin DY, Lee LY, Wang Ch-W (2006) Adv Drug Deliv Rev 58:1274–1325
8. Barat A, Ruskin HJ, Crane M (2008) Theor Biosci 127:1611–7530
9. Zhou Y, Wu XY (2002) J Control Release 84:1–13
10. Higuchi T (1961) J Pharm Sci 50:874–875
11. Higuchi T (1963) J Pharm Sci 52:1145–1149
12. Bonny JD, Leuenberger H (1991) Acta Pharm Helv 66:5–6
13. Laaksonen TJ, Laaksonen HM, Hirvonen JT, Murtomaki L (2009) Biomaterials 30:1978–1987
14. Liao Y, Lucas D (2009) Chem Eng Sci 64:3389–3406
15. Dluska E, Markowska-Radomska A (2010) Chem Eng Technol 33:1471–1480
16. Dluska E, Markowska A (2009) Chem Eng Process 48:438–445
17. Dluska E, Markowska-Radomska A (2010) Chem Eng Technol 33:113–120
18. Cortesi R, Nastruzzi C, Davis S (1998) Biomaterials 19:1641–1649
19. Kataoka T, Nishiki T, Kimura S (1989) J Membr Sci 41:197–209
20. Hino T, Kawashima Y, Shimabayashi S (2000) Adv Drug Deliv Rev 45:27–45
21. Lian G, Malone ME, Homan JE, Norton IT (2004) J Control Release 98:139–155

Progr Colloid Polym Sci (2012) 139
DOI: 10.1007/978-3-642-28974-3_7

Insights into Catanionic Vesicles Thermal Transition by NMR Spectroscopy

Gesmi Milcovich and Fioretta Asaro

Abstract Oppositely charged ionic surfactants can self-assemble into hollow structures, called catanionic vesicles, where the anionic-cationic surfactant pair assumes a double-tailed zwitterionic attitude. In the present work, multilamellar-to-unilamellar thermal transition of a mixed aqueous system of sodium dodecyl sulphate (SDS) and cetyl trimethyl ammonium bromide (CTAB), with a slight excess of the anionic one, has been investigated by ^{1}H, ^{2}H, ^{14}N NMR spectra and ^{23}Na transverse relaxation measurements. It has been inferred that an increase of the temperature enhances the SDS counterion dissociation, which can be considered as one of the driving forces of the mentioned transition. Moreover, interesting ^{23}Na T_2 changes with temperature have been detected for unilamellar aggregates.

Introduction

Amphiphiles are characterized by a double behavior: one moiety of the molecule is "solvent-loving" (lyophilic), while the other is "solvent hating" (lyophobic). Two oppositely charged single-tailed surfactants can associate into a zwitterionic pseudo-double chained structure [1, 2] (Fig. 1).

SDS/CTAB mixed systems, with a slight excess of the anionic surfactant, have been studied. The latter tends to spontaneously aggregate into multi-walled vesicular structures [3]. They can be susceptible to multi-to-unilamellar transition, which is driven by different factors, like salt/co-solutes addition, chain length and temperature [4]. Herein, temperature effects have been investigated by means of ^{1}H and ^{23}Na NMR. Sodium corresponds to the counterion of the anionic component.

G. Milcovich (✉)
Department of Chemical and Pharmaceutical Sciences, University of Trieste, Via L. Giorgieri, Trieste, 1 – 34127, Italy
e-mail: gesmi.milcovich@phd.units.it

^{23}Na isotope, which possesses a natural abundance of 100% and high NMR sensitivity, has a spin of $I=3/2$; therefore, its relaxation is dominated by the quadrupolar mechanism and NMR dynamic parameters are suitable to explore self-assembly of catanionic systems [5].

Nowadays, these vesicular systems are of increasing interest as they are widely employed in pharmaceutical/biotechnological field (e.g. targeted gene therapy, medicated syrups, eye drop products, etc.) [6, 7]. Indeed, they mimic biological membranes and related compartmentalization properties, noteworthily their preparation is quite cheap and easy [1, 3]. However, studies concerning their ultimate thermodynamically-stable structure are still undergoing.

Experimental

Sodium dodecyl-sulphate (SDS) has been obtained from BDH Chemicals Ltd. Pool. England (purity grade 99.0%), while cetyl-trimethyl ammonium bromide (CTAB) has been purchased from Sigma-Aldrich (puriss ≥96%). Aqueous solutions of CTAB and SDS have been prepared and subsequently mixed in order to obtain vesicular solutions. Different molar ratios (R) have been employed, at a constant C_{TOT} =6 mM (i.e. about 0.2% wt).

$$R = \frac{[SDS]}{[CTAB]}$$

NMR measurements were carried out on a Jeol Eclipse 400 NMR spectrometer (9.4 T), equipped with a Jeol NM-EVTS3 variable temperature unit, operating at 400MHz for ^{1}H, 61.37 MHz for ^{2}H, 28.88 MHz for ^{14}N and 105.75 MHz for ^{23}Na without field frequency lock, except for ^{1}H-NMR (lock on $CDCl_3$, in coaxial tube). The ^{23}Na-R_2 (transverse relaxation rate $R_2 = 1/T_2$) were measured by Hahn Echo.

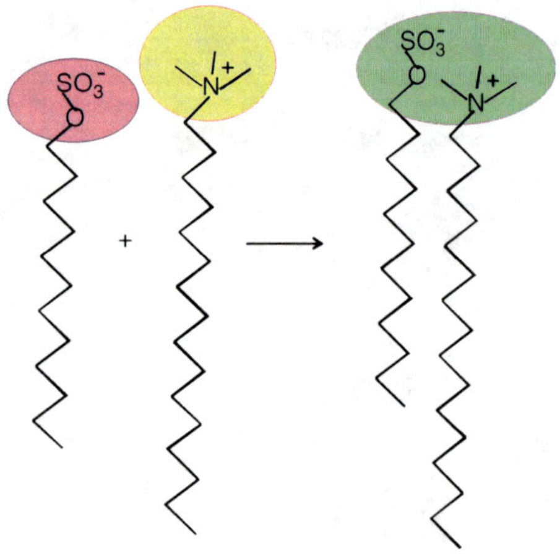

Fig. 1 Pairing of two oppositely charged single tailed amphiphiles

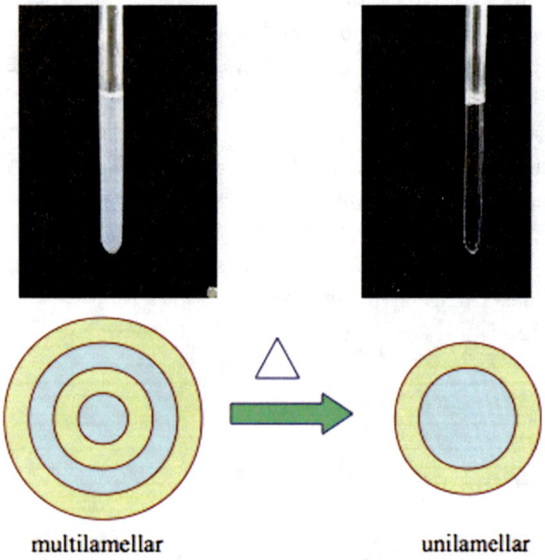

Fig. 2 Decrease of turbidity upon heating for R = 1.7 sample

Results and Discussion

Considering that a net charge allows vesicles to be stable, an excess of sodium dodecylsulphate has been used, i.e. $R > 1$ (R = SDS/CTAB molar ratio). When prepared, the vesicle solution appears milky, due to the presence of multilamellar structures [3], while immediately after the thermal transition, turbidity disappears (Fig. 2), reflecting the conversion to unilamellar aggregates with lower hydrodynamic radius [3].

The lack of water signal splitting in ^{2}H-NMR spectrum and the absence of birefringence in the polarized light microscopy observations have ruled out the presence of anisotropic lamellar phase, both prior and after thermal transition.

Relevant information into the bilayers ordering has been provided by the ^{14}N NMR spectrum (^{14}N nucleus has $I = 1$) of the tetraalkylammonium head group, which is the molecular moiety with the lowest mobility. The residual quadrupolar splitting (23.5 kHz) is comparable to those found in diluted lyotropic liquid crystals. Despite the isotropy of the sample, the line width is rather small, due to an orientational effect caused by the magnetic field, as already known for liposomes.

^{23}Na transverse relaxation rate, R_2, is much more sensitive to slow motions than longitudinal relaxation rate, R_1. Considering that catanionic systems are quite diluted, no significant deviation of the echo decay from a single exponential has been detected, neither before nor after the transition. Thus R_2 are averages of those for central and satellite transitions [8]. ^{23}Na R_2 responds to the interaction of the counterion with negatively charged aggregates, being remarkably higher than for free Na^+ (e.g. NaI 0.1 M solution), while R_1 values do not differ significantly from that of free Na^+.

Moreover, it has been noticed that larger ^{23}Na R_2 values correspond to higher R.

Upon increasing temperature, an R_2 decrease has been observed, while approaching the critical transition temperature, till it reaches free Na^+ value (Fig. 3).

The decrease of R_2 values on increasing temperature can be mainly attributed to Na^+ dissociation from the aggregates. The latter influences the packing parameter, which is responsible for the spontaneous interfacial curvature. The R_2 measured some days after the samples have been brought back to room temperature were higher (especially for the higher molar ratios), due to changes in the correlation times of the motions modulating the quadrupolar interaction. Probably it is also related to lower sodium dissociation, thus confirming an increase of dodecylsulphate content in the vesicles after the transition, as inferred from ^{1}H-NMR (Fig. 4).

Detectable ^{1}H-NMR signals correspond to those of the free anionic surfactant (no signal for cetyltrimethylammonium), which is in slow exchange with that embedded in the vesicles, as confirmed by diffusion NMR measurements. ^{1}H dodecylsulphate resonances exhibit an increase in the line width altogether with heating cycle, due to faster exchange among bulk and vesicles.

Integrals analyses showed lower values at high temperatures, related to an uptake of the anionic component into the aggregates and therefore to a greater dodecylsulphate amount in vesicles composition.

Considering that dodecylsulphate possesses the shortest chain length of the system, so the anionic excess may promote vesicle curvature, stabilizing the aggregates. Indeed, after heating transition, unilamellar form prevails.

Fig. 3 ^{23}Na-R_2 trends versus temperature for different molar ratios (R), both before and after thermal transition, compared with a NaI 0.1 M solution (free Na^+), and a micellar solution of SDS

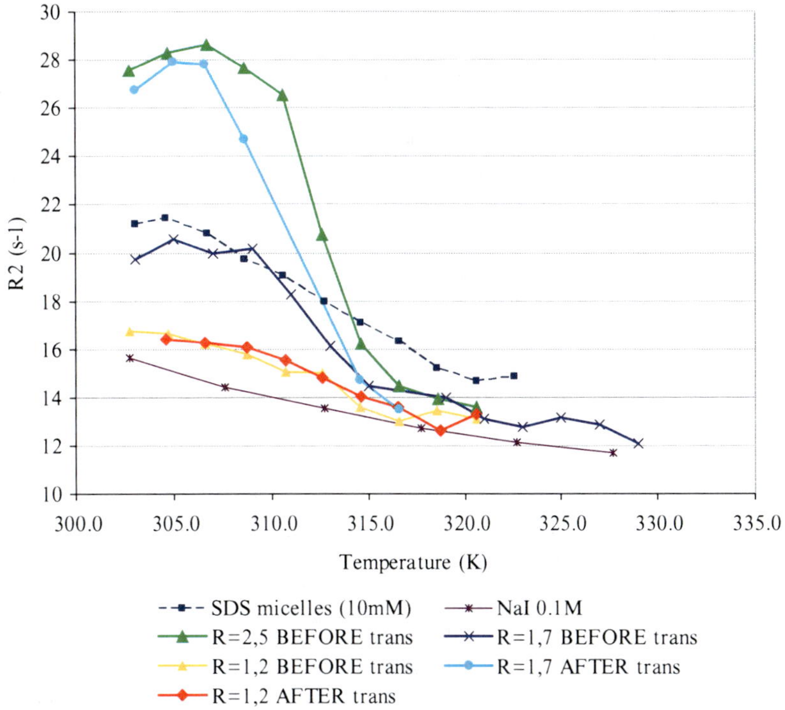

Fig. 4 ^{1}H-NMR integrals of an SDS/CTAB aqueous solution (R = 1.5). From top to bottom: 303 K, 323 K (shifted by 0.073 ppm). Peaks of water included in $CDCl_3$ are marked with *, used as integrals reference

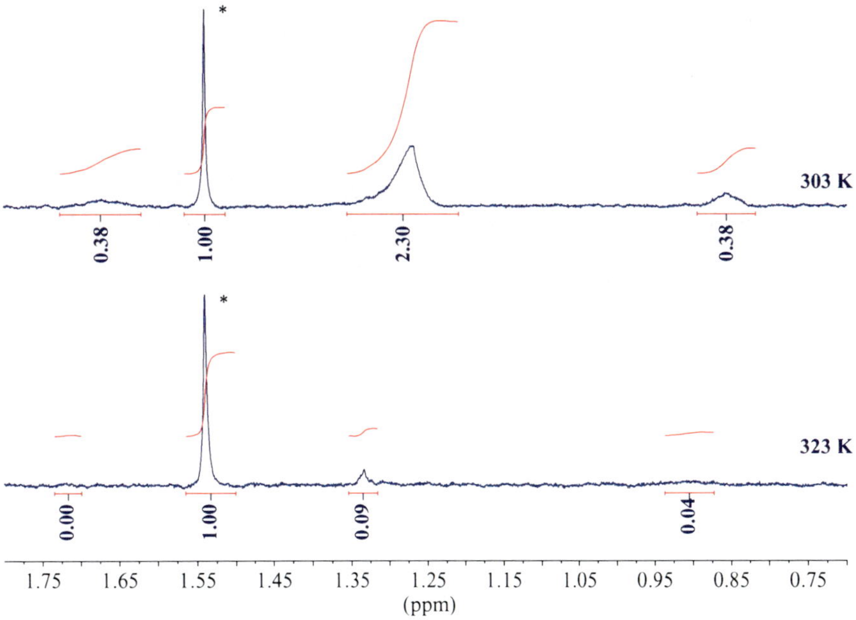

Conclusions

^{23}Na transverse relaxation rates, which are highly responsive to slow motions, display interesting trends on temperature changes. The description of the systems afforded by ^{23}Na NMR during heating evidences a dissociation of sodium, which takes place prior to the critical transition temperature [3].

At the same time, ^{1}H-NMR spectra suggest an increase of SDS/CTAB ratio in the bilayer composition.

The net charge of the surface, together with steric interactions [3], might result in a strong repulsion between bilayers, leading to a multilamellar to unilamellar transition. The latter may be stabilized by spontaneous curvature, at high temperature.

Multilamellar vesicles should be generated at room temperature, as heating can interfere with the preparation procedure, originating unilamellar structures, which are metastable, when back at room temperature.

In conclusion, catanionic vesicles showed interesting temperature tunable changes both in their lamellar composition and curvature.

Further analyses are still undergoing for what concerns the surfactant exchange between bulk and vesicular aggregates.

Acknowledgements The University of Trieste is gratefully acknowledged for financial support (FRA 2009).

References

1. Kaler EW, Murthy AK, Rodriguez BE, Zasadzinski JA (1989) Science 245:1371–1374
2. Salkar RA, Mukesh D, Samant SD, Manohar C (1998) Langmuir 14:3778–3782
3. Andreozzi P, Funari SS, La Mesa C, Mariani P, Ortore MG, Sinibaldi R, Spinozzi F (2010) J Phys Chem B 114:8056–8060
4. Antunes FE, Marques EF, Miguel MG, Lindman B (2009) Adv Colloid Interface Sci 147–148:18–35
5. Furò I (2005) J Mol Liq 117:117–137
6. Bonincontro A, Falivene M, La Mesa C, Risuleo G, Ruiz-Pena M (2008) Langmuir 24:1973–1978
7. Guida V (2010) Adv Colloid Interface Sci 161:77–88
8. Woessner DE (2001) Concepts Magn Reson 13:294–325

Progr Colloid Polym Sci (2012) 139
DOI: 10.1007/978-3-642-28974-3_8

Competitive Solvation and Chemisorption in Silver Colloidal Suspensions

Marco Pagliai[1], Francesco Muniz-Miranda[2], Vincenzo Schettino[1,2], and Maurizio Muniz-Miranda[1]

Abstract Raman spectra and ab initio computational analysis involving Car–Parrinello molecular dynamics simulations and Density Functional Theory approach have been employed to obtain information on the behaviour of oxazole and thiazole in aqueous suspensions of silver nanoparticles, where solvation and chemisorption processes competitively occur. The solvation of both oxazole and thiazole is dependent on stable hydrogen bonds with water, mainly involving the nitrogen atoms of the heterocycles. The adsorption on silver colloidal nanoparticles is, instead, ensured by replacing water molecules of the aqueous environment with surface active sites that can be modelled as Ag_3^+ clusters. These surface complexes can reproduce accurately the observed surface-enhanced Raman spectra, particularly concerning the most significant frequency-shifts with respect to the normal Raman spectra in aqueous solutions and the relative intensity changes.

Introduction

The adsorption of molecules on metal nanoparticles in colloidal suspensions represents a fundamental process for wide applications regarding the biomedical and environmental fields, as well as those related to the heterogeneous catalysis [1]. For example, nanohybrids consisting of biomolecules linked to gold or silver colloidal clusters can be employed to selectively act against pathogenic agents or tumour formations [2]; the capability of metal colloids to adsorb pollutants can overcome problems of environmental contamination [3]; the adsorption of reactants on metal nanostructured substrates like colloidal suspensions is a prerequisite for the activation of many catalytic reactions [4]. Furthermore, the aqueous dispersions of nanosized particles of coinage metals (Au, Ag, Cu) exhibit peculiar optical properties, due to the excitation of surface-localized electrons [5]. When electromagnetic radiation interacts with a metal surface with nanoscale roughness, the conduction electrons can be trapped in the nanostructures, producing collective electron excitations called surface plasmons, which lead to enhancement of the local electromagnetic field owing to the high absorption of light near the metal nanoparticles. This induces enhancements of the spectroscopic signals, like in surface-enhanced Raman scattering (SERS) [5, 6]. SERS spectroscopy allows obtaining giant Raman enhancements for molecules adsorbed on nanostructured substrates of metals like silver, gold and copper, usually around 10^6–10^7, but up to 10^{14}–10^{15} factors in single-molecule experiments. This effect is generally attributed to two different mechanisms, involving both the enhancement of the electric field near the surface, due to the resonance of the excitation wavelength with the surface plasmons of the metal nanoparticles, and the enhancement of the molecular polarizability when the ligand molecules are chemically bound to the active sites of the metal surface. Albeit this latter "chemical effect" improves the Raman enhancement only up to 10^2, with respect to the predominant role of the electromagnetic contribution, it plays a key role in the observation of the SERS spectra by strongly affecting the frequency positions and the relative intensities of the bands. As a consequence, by observing these spectroscopic features useful information can be obtained on the adsorption phenomena and the properties of molecules linked to the metal. Ag nanoparticles in aqueous colloidal suspensions are considered the most efficient substrate for enhancing the Raman signals of adsorbates with respect to other SERS-active metal platforms like thin films, crystalline islands or surfaces roughened by chemical or electrochemical treatments. Actually, silver colloids are easy to prepare and provide strong SERS enhancements of

M. Muniz-Miranda (✉)
[1]Dipartimento di Chimica "Ugo Schiff", Università di Firenze, Via della Lastruccia 3, Sesto Fiorentino, Italy+39-055-4573077
e-mail: muniz@unifi.it
[2]European Laboratory for Non-Linear Spectroscopy LENS, Via Nello Carrara 1, Sesto Fiorentino, Italy

ligands, whose adsorption can be monitored by observing in the UV-vis region the surface plasmon resonance (SPR) bands of the Ag nanoparticles. These latter show a SPR band around 390 nm, due to the electron excitation of non-aggregated particles, but it can move to longer wavelengths due to colloidal aggregation when the ligand is strongly adsorbed on metal. Actually, chemisorption of organic ligands could remove the surface charges that give stability to the colloidal suspension, inducing particle aggregation. Raman measurements, then, usually confirm this adsorption process by observing strong enhancements of the SERS bands and sizeable frequency-shifts with respect to those observed in the normal Raman spectrum.

Water, as dispersing medium of Ag nanoparticles, plays a fundamental role in both the stability of the colloids and the adsorption of organic ligands. Actually, the metal surface usually becomes negatively charged by adsorption of anionic species deriving from the aqueous environment, like hydroxide ions, which impair the aggregation and collapse of the colloidal dispersions [7]. Organic ligands, moreover, must be partially soluble in water to interact with the metal particles; since the solvation represents a necessary prerequisite for the adsorption of ligands in Ag hydrosols. Chemisorption, however, is a quite complex process, which involves not only the solvation of molecules in the aqueous environment, but also the ligand affinity to the metal and the presence of active-sites at the surface of the metal nanoparticles. As a consequence, the adsorption on silver particles in hydrosols cannot be understood if the action of the water molecules on the ligand molecules is neglected. Water largely affects the adsorption in Ag hydrosols when acid-base interactions occur between ligand molecules and aqueous medium: in this case, the SERS effect results closely dependent on the pH of the water solution; see, for instance, the cases of 1,2,3-triazole [8], tyrosine [9], 3-thiophene carboxylic acid [10], 6-mercaptopurine [11] or uracil [12]. However, also for non-protic ligands, a strong interaction with the aqueous medium could exist by effect of hydrogen bonding, when heteroatoms are present in the molecules, thus impairing chemisorption on the metal substrate. A detailed description of the H-bond dynamics is necessary to characterize the stability and strength of the ligand interaction with the solvent and to set up a suitable model to explain the Raman spectra in Ag hydrosols of molecules as oxazole and thiazole, where two different heteroatoms are present in five-membered rings (see Fig. 1). Actually, in these cases a competition occurs between solvation and chemisorption; consequently, the analysis of the SERS spectra, including frequency-shifts and relative intensities, is to be performed in comparison with the normal Raman spectra in water solutions. The presence of heteroatoms, however, is not sufficient to ensure chemisorption on silver and, consequently, a strong SERS effect, which are closely dependent on the

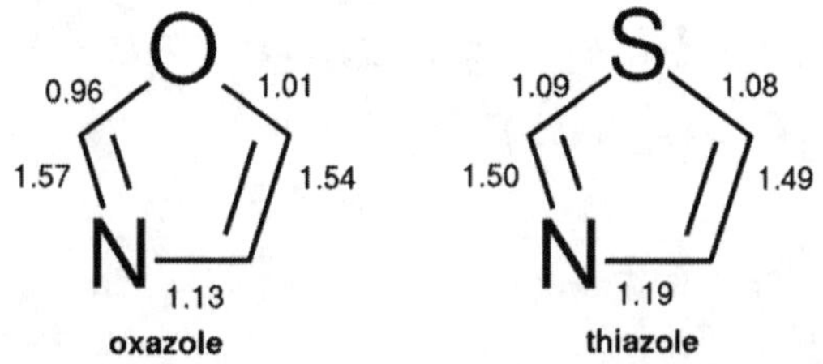

Fig. 1 Molecular structures of oxazole and thiazole, with the ring bond orders [13]

type and the number of heteroatoms, along with their position in the unsaturated ring. Hence, a detailed analysis of the interaction forces between organic and water molecules has been here performed, by using the ab initio Molecular Dynamics approach. This latter allows simplifying the model systems assumed for the subsequent DFT (density functional theory) calculations, in comparison with those performed on ligand/silver complexes able to mime the chemical interaction with the active sites of the metal surface.

Experimental

Colloidal Sample Preparation

Ag hydrosols have been prepared by adding $AgNO_3$ (99.9999% purity, Aldrich) to excess $NaBH_4$ (99.9% purity, Aldrich) following the procedure adopted by Creighton et al. [14]. One hour after the colloid preparation, LiCl was added under stirring to obtain a 10^{-3} M concentration. The ligand adsorption was obtained by adding oxazole (98% purity, Aldrich) or thiazole (99% purity, Aldrich) to silver colloids in 10^{-3} M concentration.

Absorption Spectra

UV–visible absorption spectra of the colloidal suspensions were measured with a Cary 5 spectrophotometer. The ligand adsorption was monitored by observing the red-shift of the surface plasmon resonance band, usually observed around 390 nm, due to the aggregation of the Ag nanoparticles. The addition of thiazole promotes fast aggregation of Ag nanoparticles.

XPS Measurements

X-ray photoelectron spectra were measured using a non-monochromatic Mg-K_α X-ray source (1253.6 eV) and a

VSW HAC 5,000 hemispherical electron energy analyzer operating in the constant-pass-energy mode at E_{pas}=44 eV. A few drops of the colloidal suspension were dried on glass. Then, the Ag-loaded glass was put in the UHV system under inert gas (N_2) flux and kept for 12 h. XPS spectra were referenced to C 1 s peak at 284.8 eV and the observed peaks were fitted using Gauss-Lorentz (90+10) curves after background subtraction.

Raman Spectra

Raman spectra were recorded using the 514.5 nm line of a Coherent Ar+laser, a Jobin-Yvon HG2S monochromator equipped with a cooled RCA-C31034A photomultiplier. SERS spectrum of thiazole was registered a few minutes after the ligand addition; oxazole, instead, was added during the colloid preparation to have faster adsorption and good SERS response.

Computational Details

Ab initio molecular dynamics simulations were performed with the Car-Parrinello approach [15], using the CPMD package [16] on a system made up of 64 water and one ligand molecules in the microcanonical (NVE) statistical ensemble. BLYP exchange and correlation functional was adopted, while pseudopotentials and plane waves were employed to describe the wave functions [17, 18]. DFT calculations were performed with the Gaussian suite of programs [19] using the B3LYP functional and the 6-31++G(d,p) or the TZVP basis set for all atoms, except silver (Lanl2dz basis set). The calculated frequencies were uniformly scaled by 0.98 factor. The Raman intensities of the harmonic modes corresponded to spatially averaged values according to the usual formulas reported in the literature [20–22]. Similar results were obtained for simulated Raman and SERS spectra by using the two different basis sets.

Molecular Dynamics of Oxazole and Thiazole in Water

These unsaturated five-membered heterocyclic compounds are employed as building blocks of polymers [23] or ligands in supramolecular chemistry [24] and found in many biomolecular systems [25]. Both compounds adopt a planar geometry (Fig. 1), but the different nature of oxygen and sulphur atoms gives rise to significant changes in properties like aromaticity [26–28] and basicity [29]. Here, we want to examine the behaviour of these heterocycles with respect to the water environment in Ag hydrosols. The study of the solvation dynamics of oxazole and thiazole in water with classical molecular dynamics, as performed for pyridine in Ag hydrosol [30], is complicated by the presence of two different heteroatoms, owing to the difficulty of obtaining a suitable force field for the electrostatic part [31]. Hence, this study has been here performed by ab initio Molecular Dynamics (AIMD) using the Car-Parrinello method [15].

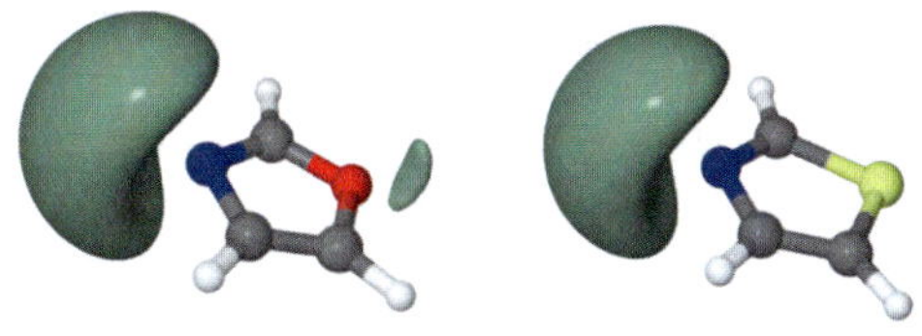

Fig. 2 Negative part of the electrostatic potential (calculated with the TZVP basis set) for oxazole (*left*) and thiazole (*right*)

For oxazole, the oxygen atom gives rise to weak interactions with the aqueous medium, as suggested also by DFT calculations, where the electrostatic potential results significantly more negative on the nitrogen atom than on the oxygen atom (Fig. 2). This agrees with the stronger H-bond ability of nitrogen in comparison to oxygen recently proposed [32] for five-membered rings and is closely related to the basicity of its lone pair. The Raman spectrum of oxazole in aqueous solution is satisfactorily reproduced by a water/oxazole complex, as shown in Fig. 3.

Thiazole interacts with the water medium only with the nitrogen atom, as expected by the electrostatic potential reported in Fig. 2. The electronic charge that can be involved in the interaction with water molecules or silver particles is concentrated on the nitrogen atom, whereas sulphur is positively-charged. Only one water molecule results H-bonded to the nitrogen atom of thiazole. The reliability of this model of interaction is validated by the good agreement between DFT-simulated and experimental Raman spectra in water solution, as shown in Fig. 3.

For oxazole the interaction with water results stronger than for thiazole; as shown by AIMD analysis of the heterocycle/water distances reported in Fig. 4, where X represents one heteroatom of the molecule. Actually, effective H-bond interactions (with $r(X \cdots H) < 2.5$ Å) are obtained for oxazole mainly with the nitrogen atom but also with the oxygen atom, whereas for thiazole only with nitrogen.

Characterization of Ag Nanoparticles in Aqueous Colloidal Suspensions

The Ag colloids used in the present investigation have been prepared by reduction of silver nitrate with sodium borohydride, following the procedure of Creighton et al. [14].

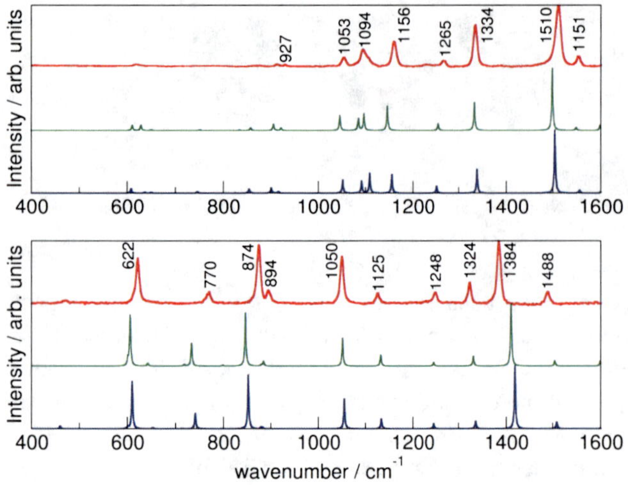

Fig. 3 Observed (*red*) and calculated (*green*: TZVP; *blue*: 6-31++G(d, p)) Raman spectra of oxazole (*upper panel*) and thiazole (*lower panel*) in water. Intensities normalized to the strongest Raman band

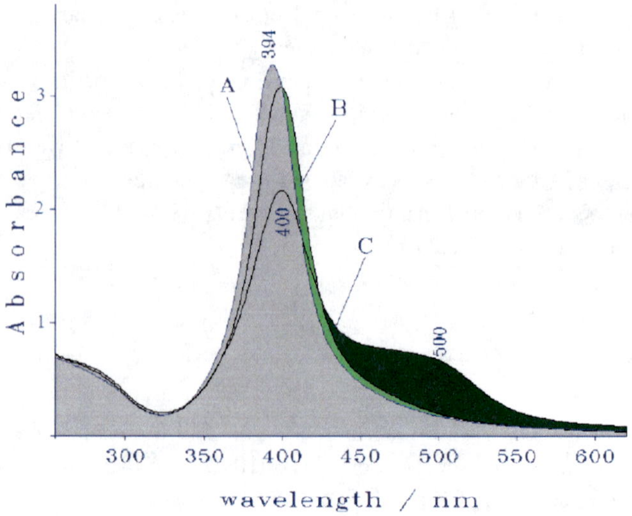

Fig. 5 Absorption spectra of pure Ag colloid (**a**) and Ag colloids 10′ after the addition of oxazole (**b**) and thiazole (**c**)

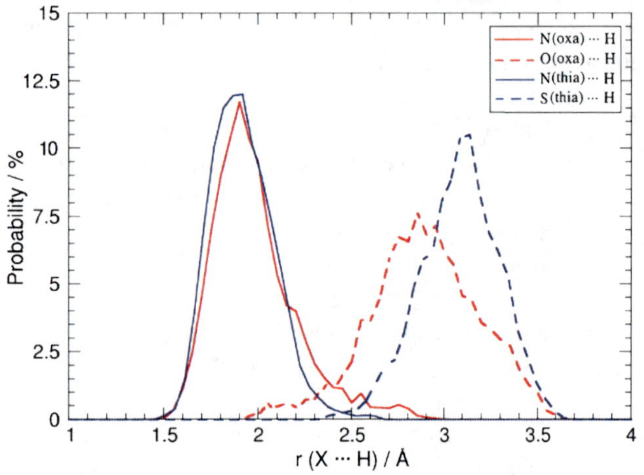

Fig. 4 Probability of interactions of oxazole and thiazole with water molecules, depending on the heteroatom (X) – H (water) distances, obtained by AIMD calculations

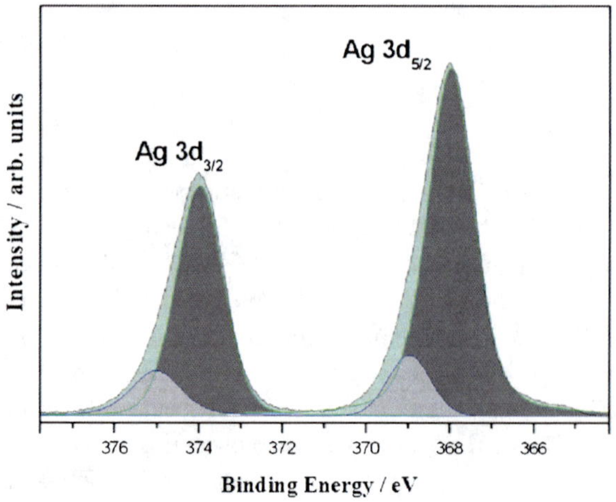

Fig. 6 XPS spectrum of Ag particles deposited on glass

The addition of LiCl to the silver colloids ensures SERS activity and colloidal stability. These properties can be deduced by observing the strong and narrow SPR band in the UV-visible absorption spectrum (Fig. 5) around 394 nm, due to non aggregated silver nanoparticles. When organic ligands are added to these Ag colloids, the chemisorption process can be revealed by the particle aggregation, which is related to the moving of the SPR band to longer wavelengths. As shown in Fig. 5, the chemisorption of thiazole induces the formation of a secondary SPR band around 500 nm, whereas oxazole appears to interact weaker with silver.

XPS (X-Ray photoemission spectroscopy) measurements have been performed in order to have information on the oxidation state of the metal nanoparticles (see Fig. 6). A sizeable amount of oxidized silver is present on the surface of the silver particles, evidenced by the presence of peaks at 369 and 375 eV related to Ag(+1), distinct from the stronger peaks at 368 and 374 eV, belonging to Ag(0).

Adsorption of Oxazole and Thiazole in Ag Colloids

The SERS spectra of oxazole and thiazole in Ag hydrosols can be satisfactory simulated by DFT calculations, adopting a model system made up of heterocyclic molecules bound to small positively-charged silver clusters modelled as Ag_3^+. The actual existence of these clusters, which act as surface active sites for the chemisorption of ligands, was ascertained in silver colloidal dispersions [33]. On the other part, this

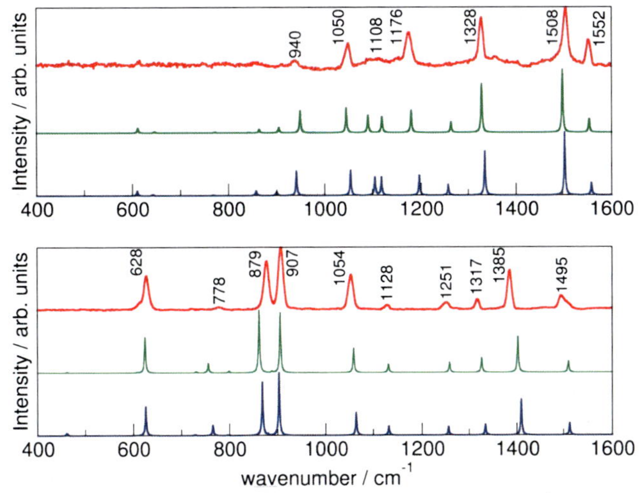

Fig. 7 Observed (*red*) and calculated (*green*: TZVP; *blue*: 6-31++G(d,p)) SERS spectra of oxazole (*upper panel*) and thiazole (*lower panel*). Intensities normalized to the strongest Raman band

Table 1 Observed and calculated (B3LYP/TZVP) Raman frequencies

Oxazole					Thiazole				
Water		Ag colloid			Water		Ag colloid		
Obs	Calc	Obs	Calc	Assignment	Obs	Calc	Obs	Calc	Assignment
927	922	940	949	H wag	622	606	628	625	C-S stret
1,094	1,097	1,108	1,119	Ring breath	894	884	907	905	Ring bend
1,156	1,146	1,176	1,179	C-O stret/H bend	1,050	1,051	1,054	1,059	Ring breath
1,334	1,333	1,328	1,328	H bend	1,322	1,331	1,317	1,327	H bend

model is consistent with the XPS results, which evidence the presence of oxidized silver on the surface of the colloidal nanoparticles, as shown in Fig. 6. The adopted model system, in particular, is expected to correctly reproduce the frequency-shifts of the SERS bands with respect to those observed in aqueous solution and their relative intensities. These latter, in fact, are strongly dependent on the enhancement effect related to the formation of chemical bonds with the metal, as stated by Otto [34]. The adsorption process of oxazole onto the silver colloidal substrate occurs by replacing the water molecule bound to the nitrogen atom of the heterocycle with the positively-charged active site of the metal surface, as argued from the molecule-metal charge transfer that results larger than in the interaction with water, 0.088 and 0.018*e*, respectively, with TZVP basis set. The DFT-simulated SERS spectrum of oxazole (see Fig. 7) accurately reproduces the experimental one, by considering both band frequencies and intensities, along with the frequency-shifts with respect to the normal Raman bands in aqueous solution, as reported in Table 1.

On the basis of the DFT calculations performed for a model system of thiazole bound to Ag_3^+ cluster, the chemisorption in Ag hydrosol is explained by the ligand/metal complex stabilization, compared to H-bond interaction with the solvent. The water molecule bound to the nitrogen atom is replaced by the surface active site, as the thiazole-silver charge transfer results larger than in the thiazole-water interaction, 0.091 and 0.018*e*, respectively, with TZVP basis set. Thiazole is bound to silver only through the nitrogen atom, whereas the sulphur atom is not involved in the interaction with the metal surface or in the H-bond interaction with water. As well as for oxazole, the DFT calculations accurately reproduce the SERS spectrum of thiazole (see Fig. 7) and the observed frequency-shifts with respect to the aqueous solution (see Table 1).

For both heterocycles the electronic charge transfer to silver clusters results similar, as well as the Ag-N bond distances (around 2.25 Å) and the corresponding binding energies (E = −124.12 kJ/mol for oxazole/silver; E = −128.30 kJ/mol for thiazole/silver, adopting the TZVP basis set). Hence, the stronger tendency of thiazole to chemisorb than oxazole, suggested by the occurrence of a secondary SPR band in the UV-visible absorption, could be mainly attributed to the kinetic mechanism of replacing water molecules, which are bonded to thiazole in a weaker way, as noted by the AIMD analysis.

As a conclusion, SERS spectroscopy, when combined with ab initio calculations including AIMD and DFT approaches, provides fundamental and accurate information on the solvation and adsorption processes of heterocyclic ligands that competitively occur in Ag hydrosols.

Acknowledgment The authors gratefully thank the Italian Ministero dell'Università e Ricerca (MIUR) for the financial support and Dr. Stefano Caporali for his help in the XPS measurements.

References

1. Schlücker S (2010) Surface enhanced Raman spectroscopy: analytical, biophysical and life science applications. Wiley-VCH, Chichester
2. Wang X, Qian X, Beitler JJ, Chen ZG, Khuri FR, Lewis MM, Shin HJC, Nie S, Shin DM (2011) Cancer Res 71:1526
3. Zhang L, Fang M (2010) Nano Today 5:128
4. Huang D, Bai X, Zheng L (2011) J Phys Chem C 115:14641
5. Le Ru EC, Etchegoin PG (2009) Principles of surface-enhanced Raman spectroscopy and related plasmonic effects. Elsevier, Oxford
6. Aroca R (2006) Surface-enhanced vibrational spectroscopy. Wiley, Chichester
7. Muniz-Miranda M, Pergolese B, Bigotto A, Giusti A (2007) J Coll Interf Sci 314:540
8. Pergolese B, Muniz-Miranda M, Bigotto A (2005) J Phys Chem B 109:9665
9. Ojha AK (2007) Chem Phys Lett 340:69
10. Chandra S, Chowdhury J, Ghosh M, Talapatra GB (2011) J Phys Chem C 115:14309
11. Szeghalmi AV, Leopold L, Pinzaru S, Chis V, Silaghi-Dumitrescu I, Schmitt M, Popp J, Kiefer W (2005) J Mol Struct 735–736:103
12. Giese B, McNaughton D (2002) J Phys Chem B 106:1461
13. Doerksen RJ, Thakkar AJ (2002) Int J Quantum Chem 90:534
14. Creighton JA, Blatchford CG, Albrecht MG (1979) J Chem Soc Faraday Trans 2 75:790
15. Car R, Parrinello M (1985) Phys Rev Lett 55:2471
16. CPMD, Copyright MPI für Festkörperforschung Stuttgart 1997–2001, Copyright IBM Corp 1990–2008
17. Pagliai M, Muniz-Miranda M, Cardini G, Schettino V (2009) J Phys Chem A 113:15198
18. Muniz-Miranda M, Pagliai M, Muniz-Miranda F, Schettino V (2011) Chem Commun 47:3138
19. Frisch MJ et al (2004) Gaussian 03 revision C.02. Gaussian, Wallingford
20. Keresztury G, Holly S, Varga J, Besenyei G, Wang AY, Durig JR (1993) Spectrochim Acta A 49:2007
21. Keresztury G (2002) In: Chalmers JM, Griffiths PR (eds) Handbook of vibrational spectroscopy, vol 1. Wiley, Chichester, p 71
22. Krishnakumar V, Keresztury G, Sundius T, Ramasamy R (2004) J Mol Struct 702:9
23. Yamamoto T, Namekawa K, Yamaguchi I, Koizumi T-A (2007) Polymer 48:2331
24. Poyatos M, Maisse-Francois A, Bellemin-Laponnaz S, Gade LH (2006) Organometallics 25:2641
25. Shatursky OYa, Volkova TM, Romanenko OV, Himmelreich NH, Grishn EV (2007) Biochim Biophys Acta 1768:207
26. Bird CW (1992) Tetrahedron 48:335
27. Mrozek A, Karolak-Wojciechowska J, Amiel P, Barbe J (2000) J Mol Struct 524:151
28. Cyrañski MK, Krygowski TM, Katritzky AR, von Ragué Schleyer P (2002) J Org Chem 67:1333
29. Kurita Y, Takayama C (1997) J Phys Chem A 101:5593
30. Pagliai M, Bellucci L, Muniz-Miranda M, Cardini G, Schettino V (2006) Phys Chem Chem Phys 8:171
31. McDonald NA, Jorgensen WL (1998) J Phys Chem B 102:8049
32. Kaur D, Khanna S (2011) Comput Theor Chem 963:71
33. Xiong Y, Washio I, Chen J, Sadilek M, Xia Y (2007) Angew Chem Int Ed 119:5005
34. Otto A (2005) J Raman Spectrosc 36:497

Progr Colloid Polym Sci (2012) 139
DOI: 10.1007/978-3-642-28974-3_9

Colloid Flow Control in *Microchannels* and Detection by Laser Scattering

Stefano Pagliara[1], Catalin Chimerel[1], Dirk G.A.L. Aarts[2], Richard Langford[1], and Ulrich F. Keyser[1]

Abstract We introduce a new approach towards the flow control and detection of colloids in microfluidic specimens. We fabricate hybrid polydimethylsiloxane (PDMS)/glass microfluidic chips equipped with parallel micrometer and sub-micrometer channels with different width and thickness. We image and detect the colloid flow direction through the microchannels by coupling laser-light-scattering in a restricted region of a single channel. We control single polymer colloids by means of a computerized pressure-based flow control system and study the Poiseuille flow through channels with different square cross section. We demonstrate the possibility of in situ sensing populations of colloids with different dimensions down to the sub-100 nm scale.

Introduction

Single particle flow control, counting and sizing in fluidic specimens is of paramount importance in environmental, industrial and clinical analysis, on-chip particle synthesis and biological sciences [1]. Micro- and nano-fluidics [2] are emerging technologies that rely on biocompatible and low cost materials and mass production fabrication processes, allow the exploitation of tiny liquid volumes and low analyte concentrations and offer an accurately controllable environment. The most common approach regarding the fabrication of *microfluidic* devices consists of a combination of photo- and soft lithography [3] that generally allows only a 2-dimensional control of the features on a same chip.

Among other technologies for the fabrication of features with variable size on a same chip – such as laser micromachining [4], electron beam and photolithography [5], multilayer soft lithography [6], gray-tone lithography [7], stereolithography [8], solid-object printing [9] and template assisted molding [10], focused ion beam (FIB) has a number of advantages such as high sensitivity and direct fabrication in selective areas without any etch mask. FIB milling has been previously exploited for the fabrication of nanofluidic channels [11] and microfluidic devices [12–14].

On the other hand among other single particle detection approaches – such as Coulter counter with nanocapillaries [15], electrical impedance [16], laser-induced fluorescence [17], particle tracking [18] and correlation spectroscopy [19], laser-light-scattering is a well established detection technique that offers a non-invasive tool for the counting of micro- and nano-particles down to the 100 nm scale such as polymer colloids, blood cells and viruses [20–23].

Here we introduce a novel approach toward the control of single sub-micrometer colloids. We fabricate microfluidic devices equipped with parallel channels with different cross section by exploiting *Platinum* (*Pt*) wires deposited via FIB as templates for soft lithography. We characterize translocations of single particles with size in the range 50–450 nm in terms of event frequency, duration and amplitude by coupling laser scattering in a single channel. *We use channels with different cross section on the same chip to investigate the pressure-driven transport of single polymer colloids with diameter of* 300 *nm*. We demonstrate the in situ sensing of populations of colloids with diameters between 50 and 450 nm.

Experimental

Preparation of Colloidal Suspensions

As test particles for our setup we used polystyrene (PS) nanospheres with mean diameter (52 ± 8)nm and (457 ± 11)nm (Polysciences, Inc. Warrington, PA) in a 2.67% and

S. Pagliara (✉)
[1]Cavendish Laboratory, University of Cambridge, Cambridge, UK
e-mail: sp608@cam.ac.uk
[2]Department of Chemistry, Physical and Theoretical Chemistry Laboratory, University of Oxford, Oxford, UK

2.63% solids (w/v) aqueous suspension, respectively. In addition poly(methyl methacrylate) (PMMA) nanospheres *with mean diameter of* (290±55) are synthesized by means of emulsion polymerization [24, 25] and dispersed in a 2.63% solids (w/v) aqueous suspension. The saline buffer for the colloidal suspensions is a KCl solution with molarity in the range 5–50 mM.

Chip Fabrication

The fabrication of the microfluidic chip consists of three steps: [26] *(a) the deposition of Pt wires on a Silicon substrate via FIB followed by (b) the patterning of a photoresist layer via photolithography for the realization of a reusable mold; (c) the replica molding of the latter in PDMS and the chemical bonding on a glass substrate via oxygen plasma functionalization for the fabrication of the final disposable device*. The FIB assisted deposition of the Pt wires is carried out with a Cross-beam 1540 FIB/SEM system (Zeiss, Oberkochen, Germany) equipped with a Ga+beam. A typical Pt deposition is carried out by using an accelerating voltage of 30 kV and a beam current of 100 pA. The scanning frequencies are 20,000 and 200 Hz along the longitudinal and orthogonal wire axis, respectively. For the fabrication of the mold, a layer of AZ 9260 (Microchemicals GmbH, Ulm, Germany) is deposited via spin coating (2,000 rpm for 30 s) on the silicon print master previously cleaned by sonication in acetone and isopropyl alcohol. After a 3 min pre-bake step at 115°C to remove the residual solvent, the sample is exposed to UV light (365–405 nm, 52 mW/cm^2) through a quartz mask (Photodata Ltd, Hitchin, UK) selectively coated with a thin Chromium film *patterned with two symmetrical stirrup shapes separated by a* 18 μm *gap* (Fig. 1a) and ending with four 2 mm-side square pads. Sample and mask are carefully aligned through a MJB4 mask aligner (Karl Suss, Garching, Germany) in a way that the central region of the wire array is positioned under the 18 μm-gap on the mask (Fig. 1b). The sample is exposed for 10 s in hard contact mode (by realizing a vacuum around 0.8 Bar between sample and mask), developed in a deionized water solution of AZ 400k developer (4:1 in volume) for 8 min at steps of 2 min each and finally rinsed with deionized water and dried with nitrogen. The thickness of the obtained photoresist structures deposited over the Pt wires (Fig. 1a) is around 12 μm as measured by a Dektak stylus profilometer (Veeco, Plainview, NY). The sample is baked at 60°C for 3 h and left in air overnight to allow complete evaporation of the solvent.

Replica molding of the device is realized by casting on it a 9:1 (base:curing agent) PDMS mixture and in situ curing at 60°C for 40 min in oven. A typical device is shown in the SEM micrograph of Fig. 1c with the inlet and outlet reservoirs separated by a 18 μm-wide and 12 μm-thick PDMS wall and connected through three hollow channels with square cross section of 1, 4 and 9 μm^2 (from left to right, respectively). Four 1.5 mm-wide circular holes are drilled by a 1.5 mm-wide circular disposable biopsy punch (Kai Industries Co. Ltd., Seki City, Japan) in correspondence of the four square pads to enable fluidic access to the microchannels. PDMS is bonded to a glass slide by exposing both surfaces to oxygen plasma treatment (8.5 s exposure to 2.5 W plasma power, Plasma etcher, Diener, Royal Oak, MI). 1.6 mm-wide PEEK tubing (Kinesis, St Neots, UK) is integrated in the holes exploiting the PDMS flexibility thus ensuring tight and fully sealed connections. The device is completed by the connection to external PEEK shut off valves (0.020″ thru-hole, 1/16″ Fittings, Kinesis) on their turn connected to a computerized pressure-based flow control system (maximum applied pressure 75 mbar, sub-mbar pressure steps MFCS-4C, Fluigent, Paris, France) that allows to stop and accurately regulate the flow in the microfluidic chip. The pressure gradient is defined as positive when the pressure applied to the inlet is higher than the

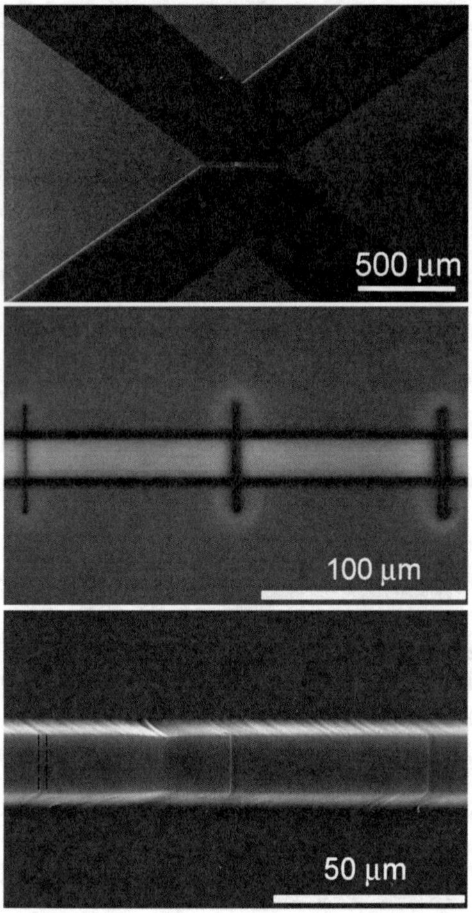

Fig. 1 (**a**) Optical micrograph of the mold: two 12 μm-*thick symmetrical* stirrups made of AZ 9260 are separated by a 18 μm gap. (**b**) Particular of the 18 μm-*long window* of uncoated Si and Pt wires with square cross section of 1, 4 and 9 μm^2. (**c**) SEM micrograph of the resulting PDMS negative replica (tilted at an angle of 38° with respect to the SEM beam column) with hollow channels connecting the inlet and outlet reservoir chambers of the microfluidic chip. *Dashed lines* mark the smallest channel

one applied to the outlet. In the same way the translocation frequency through the channels is defined as positive when the colloids flow from the inlet to the outlet.

Detection Set-up

Details about the laser scattering detection set-up are reported elsewhere [26]. Briefly a red laser beam is coupled into an oil immersion objective and thus focused in a single microchannel. The scattered laser-light is coupled to a four quadrant photodiode, the voltage signal is amplified and digitized. Pressure-driven translocations of colloids appear as increases in the voltage trace and are isolated by using a custom-made program (LabVIEW 8.6, National Instruments). Specifically the background signal or baseline is calculated every 1,000 points of the trace. The translocation events are recorded when a consecutive number of points (i.e. 40) exceed twice the value of the baseline standard deviation. Translocations of 300 and 50 nm particles through a microchannel with square cross section of 1.4 μm^2 are reported in Fig. 2a and b, respectively. Each single isolated event is fitted by a Gaussian curve (solid lines in Fig. 2c and d, respectively) which allows determination of the duration and amplitude of the translocation event (horizontal and vertical arrows, respectively, in Fig. 2b and d), the time difference with the previous event and the *signal/noise (S/N) ratio. It is noteworthy to observe that the average measured signal/noise ratio and the time difference between two successive events decreases from* 60 *to* 4 *and from* 350 *to* 50 ms, *respectively, for particle diameters of* 300 *and* 50 nm. *The lower S/N is due to the decreased amplitude in the signal that is reflected back into the objective from the smaller particle surface, while the shorter interval within two successive events is due to the presence of a higher number of particles,* 10^{11} *and* 4×10^{8} cm^{-3} *for* 50 *and* 300 nm *populations, respectively.*

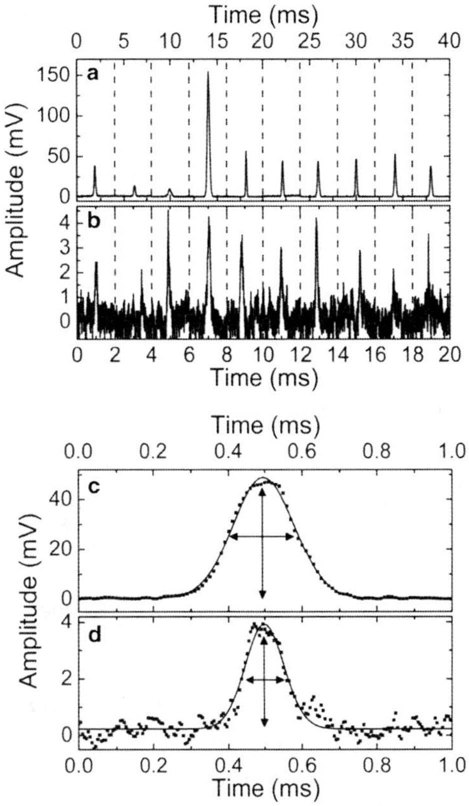

Fig. 2 *Selected intervals of* 4 (**a**) *and* 2 ms (**b**) *showing ten single events isolated from the voltage traces of* 300 (**a**) *and* 50 nm *colloids* (**b**) *translocating a microchannel with square cross section of* 1.4 μm^2. Measured signal of a single event (*squares*) fitted by a Gaussian curve (*solid lines*) with the estimation of translocation duration and amplitude (*horizontal* and *vertical arrows*) for 300 nm (**c**) and 50 nm (**d**) colloids

Results

We investigated the pressure-driven colloid transport through channels with different square cross section. We carried out experiments on two different devices each equipped with an array of three channels with cross section of 0.4, 1.4, 3.2 μm^2 and 1, 4, 9 μm^2, respectively. We measure *the translocation frequency of* 300 nm *particles* with respect to the channel cross section (Table 1) under an applied pressure gradient of 40 mbar. *The general description of hydrodynamic phenomena by the Navier-Stokes equation reduces in most of the microfluidic systems to the linear Stokes equation and the so called Stokes or creeping flow since in the limit of low Reynolds numbers the non-linear term can be neglected* [27]. *In particular the particle flow in closed channel systems can be described by introducing the channel and particle Reynolds numbers* (Table 1), Re_c and Re_p: [28]

$$\mathrm{Re}_c = \frac{U_m\sqrt{S}\rho}{\eta}, \ \mathrm{Re}_p = \frac{U_m d^2 \rho}{\eta\sqrt{S}} \tag{1}$$

where U_m *is the maximum velocity of the channel flow,* S *is the channel cross section,* d *is the particle diameter,* η *and* ρ *the dynamic viscosity and density of the flowing solution. The small values of the channel Reynolds number* (Table 1) *indicate that the non-linear term is negligible.* In particular the pressure-driven, steady-state flow through long, straight and rigid microchannels with square cross section can be described by the Hagen-Poiseuille flow that predicts a second power law of the volumetric flow rate, Q, with respect to S: [27]

$$Q \approx 0.27\frac{\Delta p}{12\eta L}S^2 \tag{2}$$

Table 1 Microchannel cross section S, measured f_m and predicted f_p translocation frequencies and corresponding errors. The error in S is evaluated by considering a 100 nm uncertainty in the SEM measurement of the channel width and height. The errors in f_m is the standard deviation calculated by averaging over measurements acquired for an interval of 30 s. For the error in f_p we take into account the error in S and a ~100 nm uncertainty in the diameter of the laser spot. The values refer to experiments with 300 nm particles

S (μm^2)	Re_c	Re_p	ω (s^{-1})	f_m (s)	f_p (s)
0.4±0.1	0.001	3.1*10^{-4}	2*10^3	0.5±0.1	0.3±0.2
1±0.2	0.007	6.1*10^{-4}	3*10^3	0.9±0.3	2.5 ±1
1.4±0.2	0.012	7.7*10^{-4}	4*10^3	3 ±0.4	4.3±1.6
3.2±0.4	0.040	1.1*10^{-3}	6*10^3	5.2±0.6	6.5 ±2.2
4±0.4	0.055	1.2*10^{-3}	7*10^3	8.6±1	7.2 ±2.4
9±0.6	0.184	1.8*10^{-3}	10^4	9.3±0.9	10.8±3.4

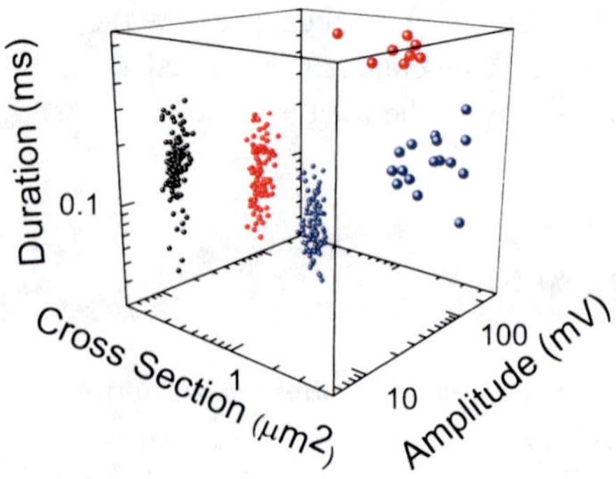

Fig. 3 3D scatter plot reporting duration and amplitude of 100 translocations of 50 and 450 nm colloids (*small* and *large spheres*, respectively) through channels with square cross section of 0.4, 1.4, 3.2 μm^2 (*black*, *red* and *blue spheres*, respectively)

where L is the channel length and Δp is the pressure gradient. For a square cross section the error of this approximate result is around 13% [27]. *At the connection between the reservoirs and the microchannels the Poiseuille description is still approximately correct since* Re_c *remains* ≤ 1 *which means that the non-linear term in the Navier-Stokes equation has a vanishingly small contribution and that the inertia effect at the microchannel inlet are negligible. Moreover the laser detection was coupled in the central part of each microchannel far away from the channel inlet and outlet. For particles dispersed in low concentration dispersion, particle flow is described by the fluid flow. In fact, since the particle Reynolds number* (Table 1) *is small particle flow is dominated by viscous drag of the fluid* [27, 28]. Therefore the predicted translocation frequency can be described as:

$$f_p = nQ = n\frac{\Delta p}{12\eta L}0.37S^2 \quad (3)$$

where n is the number of particles per unit volume, which is estimated to be 4.2×10^8 cm^{-3}.

Moreover inertial effects such as the lift and drag force play a negligible role and the flow remains laminar [28, 29]. *It is noteworthy to observe that the flowing particles rotate following the fluid vorticity* [30]. *In fact in a Newtionian fluid in shear flow with no-slip boundary conditions imposed on the surface of the sphere, the rotation speed of a single particle,* ω, *is given by:* [30]

$$\omega = \frac{\dot{\gamma}}{2} \quad (4)$$

where $\dot{\gamma}$ *is the shear rate* [31]. *Typical rotation speed are reported in* Table 1. *No shear thickening* [32] *is expected since the suspended colloids occupy only a volume fraction down to* 10^{-6}.

For an applied pressure gradient of Δp=40 mbar and taking into account that for large channels (S>1.1 μm^2) the laser spot, a, occupies only a fraction of the microchannel volume:

$$f_p = 2.86S^2 \ \ \mathrm{Hz}/\mu\mathrm{m}^4 \text{ for S1.1 } \mu\mathrm{m}^2$$

$$f_p = 2.86\sqrt{S}\frac{4}{3}\pi a^3 \ \ \mathrm{Hz}/\mu\mathrm{m}^4 \text{ for S1.1 } \mu\mathrm{m}^2 \quad (5)$$

The translocation frequency, f_p, predicted by (5) reproduces the measured translocation frequency, f_m, within the error bars (Table 1).

Therefore the laser scattering set-up coupled into a single channel provides quantitative information about the transport of particles with diameters of a few hundreds of nanometers. Moreover the presented platform allows the sensing of particles with diameter down to the 50 nm *scale*. In particular the sensing of particles over a range of diameters is easily achieved by exploiting channels with different cross section on the same microfluidic chip. As a proof of concept 50 nm particles are initially injected in the chip and detected in three different channels with cross section 0.4, 1.4, 3.2 μm^2. Thereafter a small amount of 450 nm particles (around 1:100 w/w ratio with respect to the 50 nm ones) is injected and the translocations of both types of particles are recorded. The biggest particles reach the outlet reservoir by going through the medium and the biggest channels (red and blue large spheres) but do not travel across the smallest channel as highlighted in the scatter plot in Fig. 3. In fact both small (amplitude< 10 mV) and big particles (amplitude>30 mV) are detected in the two former channels while only small particles (black small spheres) are detected in the latter one. Therefore by simply looking at the scattering events in different channels in the same chip one can easily detect populations of particles over a range of diameters.

The presented novel microfluidic platform can be readily exploited to investigate the interactions between the flowing particles and the device surface by studying the transport parameters (i.e. event frequency, amplitude, duration) as a function of the salt concentration. *We are currently exploring the possibility of employing such microfluidic systems to mimic the diffusion of metabolites across membrane protein pores and to investigate and model the physics of single channel transport. Further tuning of the glass/PDMS surfaces through polymer or protein coating may be required for more specific biological applications such as the investigation of DNA translocations under concentration/pH gradient or electro-phoretic/osmosis force* [15]. The exploitation of stiffer material extensively used in soft lithography for the generation of 50 nm features [33] could open the way for the realization of nanochannels for nanofluidics *while the improvement of the detection set-up for example with the integration of a high speed nanoscanning piezostage could allow the investigation of the transport of particles with diameter down to the sub-*50 nm *scale.*

Conclusions

We have proposed a simple and versatile approach for the control of sub-micrometer colloids in polymer-based lab-on-a-chip systems equipped with arrays of parallel channels with different square cross section down to 0.4 μm^2. We have coupled laser scattering in single microchannels for the in situ detection of single translocating colloids with minimum detectable particle size of 50 nm. We demonstrate that the pressure-driven transport of 300 nm particles through channels with different cross section can be modeled by a Poiseuille flow. Finally we demonstrated the sensing of particles with different diameters by exploiting channels with a range of cross sections on the same microfluidic chip.

References

1. Zhang H, Chon CH, Pan X, Li D (2009) Microfluid Nanofluid 7:739
2. Salieb-Beugelaar GB, Simone G, Arora A, Philippi A, Manz A (2010) Anal Chem 82:4848
3. Duffy DC, McDonald JC, Schueller OJA, Whitesides GM (1998) Anal Chem 70:4974
4. Wolfe DB, Ashcom JB, Hwang JC, Schaffer CB, Mazur E, Whitesides GM (2003) Adv Mater 15:62
5. Jung S-Y, Retterer ST, Collier CP (2010) Lab Chip 10:2688
6. Wu H, Odom TW, Chiu DT, Whitesides GM (2003) J Am Chem Soc 125:554
7. Chung J, Hsu W (2007) J Vac Sci Technol B 25:1671
8. Mizukami Y, Rajniak D, Rajniak A, Nishimura M (2002) Sens Actuat B-Chem 81:202
9. McDonald JC, Chabinyc ML, Metallo SJ, Anderson JR, Stroock AD, Whitesides GM (2002) Anal Chem 74:1537
10. Tu D, Pagliara S, Camposeo A, Potente G, Mele E, Cingolani R, Pisignano D (2011) Adv Funct Mater 21:1140
11. Maleki T, Mohammadi S, Ziaie B (2009) Nanotechnology 20:105302
12. Campbell LC, Wilkinson MJ, Manz A, Camilleri P, Humphreys CJ (2004) Lab Chip 4:225
13. Imre A, Ocola LE, Rich L, Klingfus J (2010) J Vac Sci Technol B 28:304
14. Wanzenboeck HD, Fischer M, Mueller S, Bertagnolli E (2004) Proc IEEE Sens 1–3:227
15. Steinbock LJ, Otto O, Chimerel C, Gornall J, Keyser UF (2010) Nano Lett 10:2493
16. Segerink LI, Sprenkels AdJ, ter Braak PM, Vermes I, van den Berg A (2010) Lab Chip 10:1018
17. Andreyev D, Arriaga EA (2007) Anal Chem 79:5474
18. Otto O, Czerwinski F, Gornall JL, Stober G, Oddershede LB, Seidel R, Keyser UF (2010) Opt Express 18:22722
19. Gadd JC, Kuyper CL, Fujimoto BS, Allen RW, Chiu DT (2008) Anal Chem 80:3450
20. Kummrow A, Theisen J, Frankowski M, Tuchscheerer A, Yildirim H, Brattke K, Schmidt M, Neukammer J (2009) Lab Chip 9:972
21. Pamme N, Koyama R, Manz A (2003) Lab Chip 3:187
22. Steen HB (2004) Cytometry A 57A:94
23. Rezenom YH, Wellman AD, Tilstra L, Medley CD, Gilman SD (2007) Analyst 132:1215
24. O'Callaghan KJ, Paine AJ, Rudin A (2007) J Appl Polym Sci 1995:58
25. Kumacheva E, Kalinina O, Lilge L (1999) Adv Mater 11:231
26. Pagliara S, Chimerel C, Aarts DGAL, Langford R, Keyser UF (2011) Lab Chip 11:3365
27. Bruus H (2008) Theoretical microfluidics. Oxford University Press, Oxford
28. Di Carlo D, Irimia D, Tompkins RG, Toner M (2007) Proc Nat Acad Sci USA 104:18892
29. Kim YW, Yoo JY (2008) J Micromech Microeng 18:065015
30. Snijkers F, D'Avino G, Maffettone PL, Greco F, Hulsen M, Vermant J (2009) J Rheol 53:459
31. Girardo S, Cingolani R, Pisignano D (2007) Anal Chem 79:5856
32. Lee YS, Wagner NJ (2003) Rheol Acta 42:199
33. Odom TW, Love JC, Wolfe DB, Paul KE, Whitesides GM (2002) Langmuir 18:5314

Progr Colloid Polym Sci (2012) 139
DOI: 10.1007/978-3-642-28974-3_10

FACS Based High Throughput Screening Systems for Gene Libraries in Double Emulsions

Radivoje Prodanovic[1,2], Raluca Ostafe[2,3], Milan Blanusa[2], and Ulrich Schwaneberg[2,3]

Abstract A flow cytometry based high throughput screening system for glucose oxidase (GOx) gene libraries in double emulsions was developed. Firstly, encapsulation of yeast cells in double emulsion was optimized by changing the ABIL EM90 concentration in light mineral oil from 2.9% to 1.5%. This enabled formation of larger water droplets and more efficient yeast cell encapsulation. Several fluorescent assays for hydrogen peroxide were tested and the 3-carboxy-7-(4′-aminophenoxy)-coumarine (APCC) oxidation by horseradish peroxidase based assay best fit the requirements of the double emulsion technology. Using an optimized substrate solution consisting of 0.5 mM APCC, 40 mM glucose and 10 U/mL of horse radish peroxidase, a referent gene library containing 10^7 yeast cells was sorted in 30 min and enriched from 1% to 15% of yeast cells expressing wt GOx.

Introduction

Directed protein evolution is a method used for improving enzyme properties in iterative cycles of diversity generation and screening [1]. In directed evolution experiments, screening is the most limiting step and systems with the highest throughput are display technologies (10^6–10^{10} variants, mostly used for evolving affinity/binding [2–4]) and in vitro compartmentalization (IVC, 10^{10} reaction compartments per mL reaction volume). In IVC technology, enzymatic reactions are performed inside aqueous microdroplets of water-in-oil-in-water emulsions [5]. The diameter of microreaction compartment droplets ranges from 0.5 to 10 μm and enables, in combination with flow cytometry, sorting at a very high speed (up to 10^7 droplets per hour) if a suitable fluorescence assay is available [6, 7]. Using polymeric detergent ABIL EM 90 cross talk between internal water droplets was prevented due to its inability to form micelles [6]. Despite its general applicability, relatively few reports and reviews [5, 8] have been published on IVC based screening systems using flow cytometer sorting of double emulsions (thiolactonase [6], galactosidase [7], glucosidase [9]). The major challenge for IVC technology is the development of fluorescence assays that reflect the activity/property which is being evolved and that incorporate the complex biochemical environment of single cell enzyme measurement in double emulsion.

Detection of hydrogen peroxide production is important for many analytical applications in medicine, process engineering and food processing [10]. Enzymes associated with peroxide production are glucose oxidase (GOx) [11], amino acid oxidase [12] and cholesterol oxidase [13]. Glucose oxidase from *Aspergillus niger* is a prominent and industrially important enzyme that is used in the food industry [11], biosensors [14] and enzymatic biofuel cells [15].

Experimental

Materials

All chemicals used were of analytical reagent grade or higher and purchased from Sigma-Aldrich Chemie (Taufkirchen, Germany) and Applichem (Darmstadt, Germany). pYES2 shuttle vector and *Saccharomyces cerevisiae* strain INVSc1 were purchased from Invitrogen (Karlsruhe, Germany). Nucleotides and all other enzymes were purchased from Fermentas (St. Leon-Rot, Germany), if not stated otherwise.

R. Prodanovic (✉)
[1]University of Belgrade, Studentski trg 12, Belgrade 11000, Serbia
[2]Jacobs University Bremen, Campus Ring 1, Bremen 28759, Germany
e-mail: rprodano@chem.bg.ac.rs
[3]RWTH Aachen University, Worringer Weg 1, Aachen 52074, Germany

Instrumentation

Fluorescence micrographs were obtained on the LSM 510 Meta inverted laser scanning microscope (Carl Zeiss GmbH, Oberkochen, Germany) using an excitation wavelength of 488 nm (argon laser) and a pinhole size of one airy unit. Image analysis was performed with the Zeiss LSM Image Browser version 3.2.0.70. (Carl Zeiss GmbH, Jena, Germany). Flow cytometry analysis and sorting was performed using the Partec CyFlowML (Münster, Germany) flow cytomer. For emulsification, the MICCRA D-1 dispenser (ART Prozess- & Labortechnik Gmbh & Co. KG, Müllheim) was used. A thermal cycler (Mastercycler gradient, Eppendorf, Hamburg, Germany) and thin-wall PCR tubes (Mμlti-Ultra tubes, 0.2 mL, Carl Roth, Germany) were used for all PCRs. The amount of DNA was quantified using a NanoDrop photometer (Thermoscientific, USA).

Synthesis

3-Carboxy-7-(4′-aminophenoxy)-coumarine (APCC) was synthesized according to the protocol Pyare et al. [16] except for the reduction from 3-Carboethoxy-7-(4′-nitrophenoxy) coumarine to 3-Carboethoxy-7-(4′-aminophenoxy) coumarine was performed with iron powder in ethanol in the presence of trace amounts of hydrochloric acid [17]. The starting substrate for APCC synthesis, 3-Carboethoxy-7-(hydroxy) coumarine, was synthesized following the protocol of Chilvers et al. [18].

Enzymatic Assays

Fluorescence based assays were performed in 96 MicroWell™ Nunclon™Δ Optical Black Flat Bottom Plates (Greiner Bio One GmbH) using a Tecan GENios microplate reader (MTX Lab Systems, Inc. Virginia, USA) in a total reaction volume of 100 μL.

2′,7′-Dichlorodihydrofluorescein (DCFH, λex 485 nm, λem 535 nm): Hydrolysis of 2,7-dihydrodichlorofluorescein-diacetate (DCFH-DA, Sigma) to 2,7-dihydro-dichlorofluorescein was performed according to the protocol by Keston and Brandt [19]. For the microtiter plate (MTP) measurements, a mix of DCFH (6.25 mM), Horse radish peroxidase (2.5 U/mL) and glucose (50 mM) in sodium phosphate buffer (25 mM, 7.4) was used.

Pentafluorobenzene-sulfonyl-2,7-dichlorofluorescein (PFBSF, λex 488 nm, λem 520 nm): PFBSF was synthesized according to the protocol by Maeda et al. [20]. For the MTP measurements, a mix of PFBSF (166 μM), glucose (55 μM) in HEPES buffer (10 mM, 7.4) was used.

Epinephrine and diphenyl ethylene diamine (Epi and DFE, λex 360 nm, λem 535 nm); For the MTP measurements, a mix of Epi (3 mM), DFE (5 mM), glucose (300 mM), myeloperoxidase MP-11 (2.72 μM) in sodium phosphate buffer (1 M, 7.4) was used.

3-Carboxy-7-(4′-aminophenoxy)-coumarine (APCC) using Horse radish peroxidase (HRP), (λex 375 nm, λem 460 nm): For the MTP measurements, a mix of APCC (500 μM), HRP (10 U/mL), glucose (40 mM if not differently stated) in PBS buffer (7.4) was used.

ABTS [21] (λex 405 nm); For the MTP measurements and agar plate assay, a mix of ABTS (4 mM), HRP (1 U/mL) and glucose (5 and 333 mM) in PBS buffer (pH 7.4) was used.

Library Construction

The referent library was made by mixing *S. cerevisiae* InvSc1 containing an empty pYES2 plasmid vector and wt-GOx-pYES2 plasmid vectors containing gene for wild type glucose oxidase. Cells were grown (24 h, 30°C, 250 rpm) after transformation in liquid SCU-glucose media, centrifuged (3,000 rpm, 25°C), resuspended in SCU-galactose media (0.8 O.D., 600 nm) and incubated 4 h (30°C, 250 rpm) [22]. Before compartmentalization, cells were washed three times in PBS buffer (pH 7.4) and mixed in 99:1 ratio (pYES2 versus wt-GOx-pYES2). Sorted libraries were plated on agar plates and the percentage of positive cells in the library was checked using the ABTS agar plate assay.

Preparation of Double Emulsion

In vitro compartmentalization of *S. cerevisiae* cells was performed using a modified protocol from Aharoni et al. [6]. Washed yeast cells, together with the components necessary for the reaction (total reaction volume 25 μL), were added to the oil phase of 250 μL 1.5% (v/v) ABIL EM90 (Tego, Germany) in light mineral oil (approx 4°C). The two phases were homogenized on ice in a 2 mL round bottom cryotube (3 min, 8,000 rpm using T18 basic ULTRA-TURRAX® from IKA, Germany). The second water phase (250 μL), containing 1.5% (w/v) carboxy-methylcellulose sodium salt (CMC) 1% (v/v) Triton X 100 in PBS, was added to the primary emulsion and homogenized on ice for 3 min at 7,000 rpm.

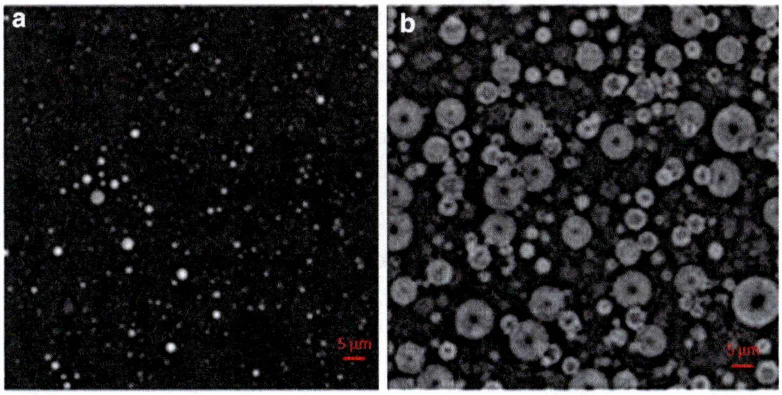

Fig. 1 Phase contrast micrograph of primary emulsion. Primary emulsion made by stirring of one volume of PBS and nine volumes of 2.9% Abil EM 90 in light mineral oil at: (**a**) 10,000 rpm, 5 min, 4°C; (**b**) 8,000 rpm, 3 min, 4°C

Flow Cytometry Sorting of Double Emulsions

Emulsions were diluted in sheat fluid (0.9% NaCl, 0.01 Triton X-100) and analyzed using a Partec CyFlowML (Münster, Germany) flow cytometer. The trigger parameter was set to forward scattering. The analysis rate was approximately 8,000 events/s and the sorting speed was 10–100/s. A blue solid state 200 mW laser at 488 nm and a 16 mW UV laser at 375 nm were used for excitation. The emission was detected with the 460 and 520 nm filter. The positive emulsion droplets were gated on the FL3-DAPI, FL1-FITC plot. The sorted cells were centrifuged, plated and grown on SCU-glucose agar plates (48 h, 30°C). GOx was induced by replica plating and growing of cells on SCU-galactose plates (24 h, 30°C). The cells were overlayed with ABTS agar containing low melting agarose (2%), ABTS (3.3 mM), HRP (5 U/mL) and glucose (333 mM) in PBS. The positive vs. negative cells were counted and the activity of the sorted library was determined in liquid culture using the ABTS detection system.

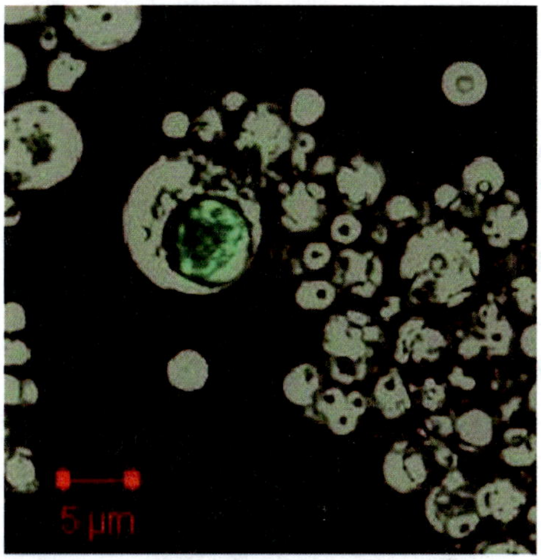

Fig. 2 Confocal epifluorescent micrograph of double emulsion with encapsulated FITC (fluorescein isothiocyanate) labeled *S. cerevisiae* cell (*green*). Double emulsion was made by stiring of one volume of primary emulsion and one volume of 1.5% (w/v) carboxy-methylcellulose sodium salt (CMC) 1% (v/v) Triton X 100 in PBS on ice, for 3 min at 7,000 rpm

Results and Discussion

The development of the screening system of yeast cells encapsulated in double emulsions was first optimized. Various, fluorescent assays were tested in microtiter plates and emulsions and HRP coupled assays were optimized with respect to glucose concentration and tested by sorting a referent GOx gene library in double emulsion with flow cytometer.

Encapsulation of Yeast Cells

The IVC technology developed for directed evolution of thiolactonase was optimized for *E. coli* cells. Double emulsions were proven to be stable to droplet coalescence and "cross-talk" between droplets, mainly due to the use of polymeric detergent ABIL EM90 that cannot form micelles [6]. Since GOx has been expressed in a active form only in yeast cells [23], which have a five times higher diameter than *E. coli* cells the published protocol [6] for in vitro compartmentalization was not suitable for encapsulation of yeast cells inside primary water droplets of double emulsion. In order to increase encapsulation efficiency of yeast cells, we adjusted emulsification protocol by decreasing the concentration of ABIL EM90 detergent to 1.5% (v/v), stirring speed to 8,000 rpm and emulsification time to 3 min for making primary emulsion. After this adjustment, diameter of primary water droplets was increased to the size of yeast cells, (Fig. 1).

Increased diameter of the internal water droplets in primary emulsion allowed us to encapsulate yeast cells inside double emulsions without changing conditions for second emulsification step (Fig. 2).

Table 1 Comparison of different fluorescent assays for peroxide detection. Fluorescent assays for GOx activity that are based on the detection of hydrogen peroxide with fluorescent probes that have absorbance at the excitation wavelengths of available lasers (375, 488 nm) and fluorescence emission fitting to filters available for the flow cytometer

Name	Reaction
2′,7′ Dichlorodihydrofluorescein (DCFH) [19]	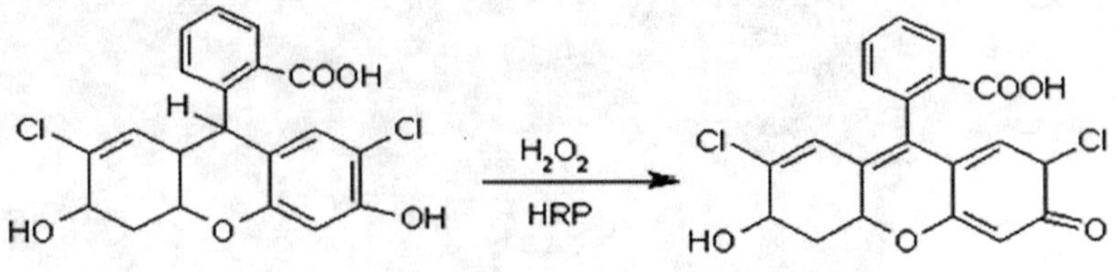 λ_{Ex} 498 nm, λ_{Em} 522 nm High fluorescence of blank reaction in emulsion already with soluble GOx. High photosensitivity of the DCFH to the irradiation light used for detection [27]
Pentafluorobenzenesulfonyl 2,7 dichloro fluorescein [20]	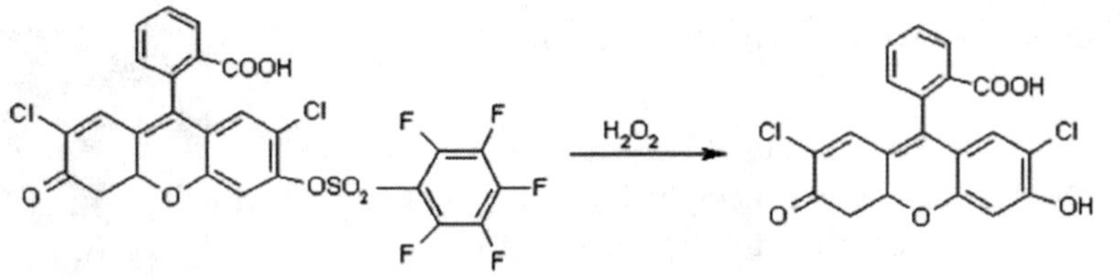 λ_{Ex} 488 nm, λ_{Em} 520 nm High fluorescence of blank reaction in emulsion where cell occupies more than half of the internal water droplet volume. PFSF due to its high hydrophobicity enters the cell and detects the endogenous peroxide
Epinephrine diphenyl ethylene diamine [24]	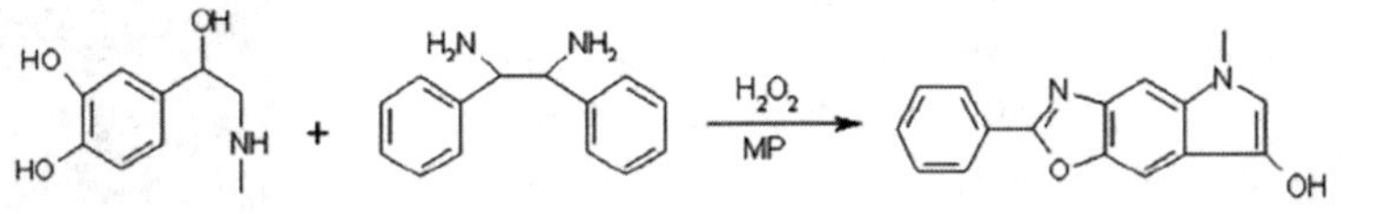 λ_{Ex} 350-360 nm, λ_{Em} 470-490 nm No fluorescence observed when reaction performed in double emulsions. Fluorescent product is diffusing out of the primary water droplets of double emulsion
3-Carboxy-7-(4′- aminophenoxy) coumarin using HRP [26]	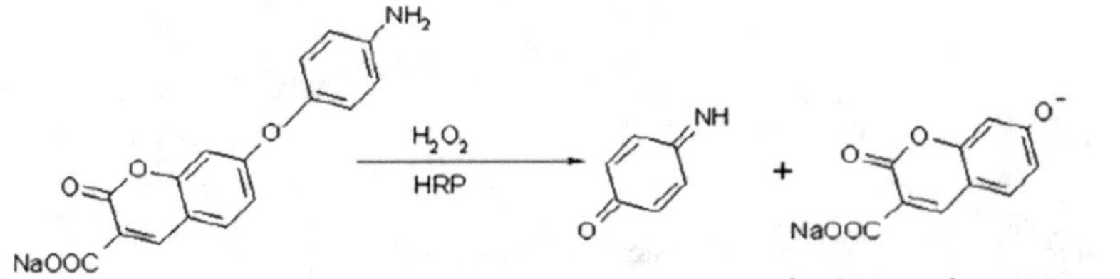 λ_{Ex} 375 nm, λ_{Em} 455 nm Fluorescent signal could be detected by flow cytometer in double emulsion and was very dependent on glucose concentration. This assay could be used for sorting of referent library made from cells not expressing GOx and cells expressing wt GOx

Using optimized protocol, 10% of the yeast cells added to the double emulsion were encapsulated inside primary water droplets. Percentage of fluorescent cells inside double emulsion droplets, was calculated by counting them under fluorescent microscope.

Testing Available Fluorescent Assays

Our aim was to develop a fluorescent assay for GOx that could be used for screening of GOx gene libraries in double emulsions for natural enzyme activity converting glucose to

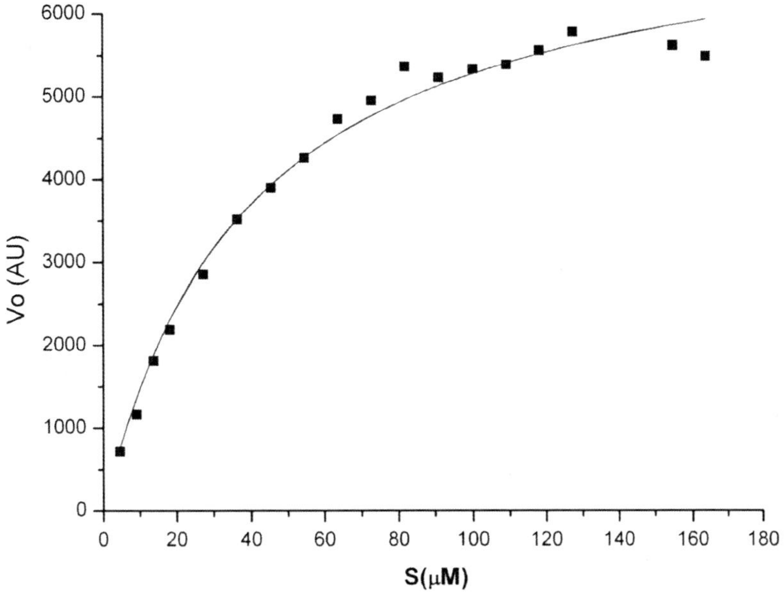

Fig. 3 Michaelis-Menten curve for oxidation of 3-carboxy-7-(4′-aminophenoxy)-coumarin (APCC) by horse radish peroxidase (HRP) at 1 mM hydrogen peroxide. Determined K_m value of HRP was 39.5 μM. Vo – increase in fluorescence signal in AU (arbitrary units) per 1 min. S – concentration of APCC

gluconic acid and thereby producing hydrogen peroxide. Hydrogen peroxide detection was chosen as a starting point for the development of the flow cytometry based screening system due to the fact that there were a large number of published fluorescent assays for hydrogen peroxide. In order to use them for flow cytometry based screening system, products of the reaction must be excited by the wavelengths of the flow cytometer lasers (375, 488 nm). Several fluorescent probes fitting the criteria above were tested for detection of hydrogen peroxide; 2′,7′-dichlorodihydrofluorescein (DCFH) [19], pentafluorobenzenesulfonyl (PFBSF) [20], epinephrine diphenyl ethylene diamine [24] and 3-carboxy-7-(4′-aminophenoxy)-coumarin [25, 26], Table 1.

It was clear from previous experiments that it is necessary to use a fluorescent probe that cannot detect peroxide inside the cell and that converts to a charged product without diffusing out of internal water droplets through the oil phase. APCC fit these criteria and was used to measure HRP activity and the detection of hydrogen peroxide [26]. APCC could not react with hydrogen peroxide inside the cell without HRP. In an enzymatic reaction that obeyed Michaelis Menten kinetics (Fig. 3), APCC yielded charged 3-carboxy-7-hydroxycoumarin (CC).

Using APCC as a fluorescent probe in MTP, a 50-fold difference in fluorescence signal between the blank reaction of cells not expressing GOx and the positive reaction of cells expressing surface GOx was obtained. There was also a detectable difference in fluorescence by flow cytometry between the blank reaction without GOx and the positive reaction with soluble GOx emulsified in double emulsion, proving the reaction product was not diffusing out through the oil phase from primary water droplets. However, in order to detect fluorescence of double emulsion droplets containing cells expressing GOx by flow cytometery, the glucose concentration was increased compared to the MTP setup (from 1 to 40 mM). The yeast cells present at high ratios of cell volume per reaction volume in emulsion water droplets (25% v/v) were absorbing limiting amounts of glucose and decreasing the fluorescence signal. After optimizing the glucose concentration, GOx activity ws detectable in yeast cells encapsulated inside double emulsion droplets, Fig. 4.

At 0 h, no fluorescent droplets were present in the blank and positive reaction, after 2 h fluorescent droplets were detected and sorted. After sorting and centrifugation, the cells were spread onto SCU-Gal agar plates for expression of GOx. After 48 h, positive colonies of yeast cells were detected on agar plates using a ABTS filter paper pre-screening method.

The SCU-GAL agar plate with the referent library before sorting contained approximately 1,000 colonies, of which 11 were positive for GOx activity (1%). The SCU-GAL agar plate with the referent library after sorting contained 28 negative colonies and 5 positive colonies with GOx activity (15% of positive in population). Similar results were obtained in the several such experiments.

From these results it can be concluded that by using FACS based screening, the referent gene library can be enriched in a single round of sorting, a 15-fold increase in cells expressing wt glucose oxidase.

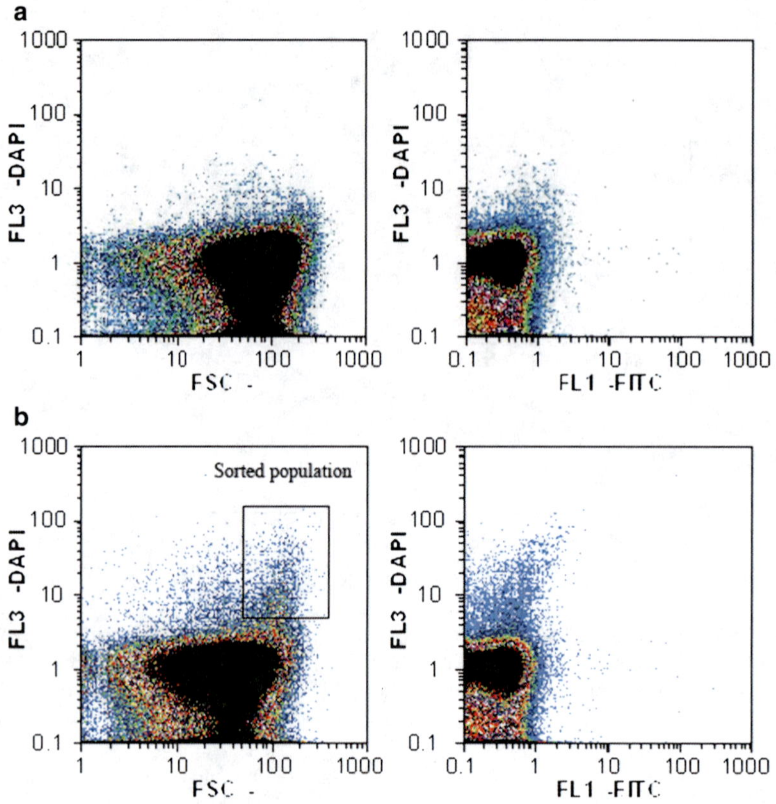

Fig. 4 Recording of referent library for GOx on FACS at 0 h (**a**) and 2 h (**b**) after emulsification. Primary water phase contained 1.25 mM L-ascorbic acid, 0.5 mM APCC, 40 mM glucose and 10 U/mL of horse radish peroxidase

Conclusions

The encapsulation of yeast cells in double emulsion was optimised and tested against several fluorescent assays for hydrogen peroxide in emulsion systems. The assay based on HRP catalysed oxidation of 4-amino-phenoxy-3-carboxy-coumarine was able to detect GOx activity by flow cytometery in double emulsions. A referent library was enriched in a single round of sorting resulted in an increase from 1% to 15% of cells expressing wt GOx. The next step is to test developed assay for sorting of error prone PCR libraries during directed evolution of GOx.

Acknowledgement The work was supported by the Alexander von Humboldt foundation, BMBF BiochancePlus program and the company BRAIN AG (Dr. Jürgen Eck and Dr. Frank Niehaus).

References

1. Matsuura T, Yomo T (2006) J Biosci Bioeng 101(6):449–456
2. Antipov E, Cho AE, Wittrup KD, Klibanov AM (2008) Proc Natl Acad Sci USA 105(46):17694–17699
3. Lipovsek D, Antipov E, Armstrong KA, Olsen MJ, Klibanov AM, Tidor B, Wittrup KD (2007) Chem Biol 14(10):1176–1185
4. Becker S, Michalczyk A, Wilhelm S, Jaeger KE, Kolmar H (2007) Chembiochem 8(8):943–949
5. Taly V, Kelly BT, Griffiths AD (2007) Chembiochem 8(3):263–272
6. Aharoni A, Amitai G, Bernath K, Magdassi S, Tawfik DS (2005) Chem Biol 12(12):1281–1289
7. Mastrobattista E, Taly V, Chanudet E, Treacy P, Kelly BT, Griffiths AD (2005) Chem Biol 12(12):1291–1300
8. Bershtein S, Tawfik DS (2008) Curr Opin Chem Biol 12(2): 151–158
9. Hardiman E, Gibbs M, Reeves R, Bergquist P (2010) Appl Biochem Biotechnol 161(1–8):301–312
10. Watt BE, Proudfoot AT, Vale JA (2004) Toxicol Rev 23 (1):51–57
11. Kirstein D, Kuhn W (1981) Lebensmittel Industrie 28(5):205–208
12. Pilone MS, Pollegioni L (2002) Biocatal Biotransform 20(3): 145–159
13. MacLachlan J, Wotherspoon ATL, Ansell RO, Brooks CJW (2000) J Steroid Biochem Mol Biol 72(5):169–195
14. Dzyadevych SV, Arkhypova VN, Soldatkin AP, El'skaya AV, Martelet C, Jaffrezic-Renault N (2008) Irbm 29(2–3):171–180
15. Liu Q, Xu XH, Ren GL, Wang W (2006) Prog Chem 18(11): 1530–1537
16. Khanna PL, SJCCC (Santa Clara), Ullman EF (Atherton, California) (1982) Method for the determination of peroxidase using fluorogenic substrates. In: Office USP (ed). vol 4857455. USA: Syntex (U.S.A) Inc., Palo Alto
17. Agrawal A, Tratnyek PG (1996) Environ Sci Technol 30(1):153–160
18. Chilvers KF, Perry JD, James AL, Reed RH (2001) J Appl Microbiol 91(6):1118–1130
19. Keston AS, Brandt R (1965) Anal Biochem 11(1):1–5
20. Maeda H, Futkuyasu Y, Yoshida S, Fukuda M, Saeki K, Matsuno H, Yamauchi Y, Yoshida K, Hirata K, Miyamoto K (2004) Angew Chem Int Ed 43(18):2389–2391
21. Zhu ZW, Momeu C, Zakhartsev M, Schwaneberg U (2006) Biosens Bioelectron 21(11):2046–2051

22. Prodanovic R, Ostafe R, Scacioc A, Schwaneberg U (2011) Comb Chem High Throughput Screen 14(1):55–60
23. Frederick KR, Tung J, Emerick RS, Masiarz FR, Chamberlain SH, Vasavada A, Rosenberg S, Chakraborty S, Schopter LM, Massey V (1990) J Biol Chem 265(7):3793–3802
24. Nohta H, Watanabe T, Nagaoka H, Ohkura Y (1991) Anal Sci 7(3): 437–441
25. Pyare LK, San Jose, Chang CC, Ullman EF (1982) United States Patent 4,857,455
26. Setsukinai K, Urano Y, Kakinuma K, Majima HJ, Nagano T (2003) J Biol Chem 278(5):3170–3175
27. Rota C, Chignell CF, Mason RP (1999) Free Radic Biol Med 27 (7–8):873–881

Progr Colloid Polym Sci (2012) 139
DOI: 10.1007/978-3-642-28974-3_11

A Dimensionless Analysis of the Effect of Material and Surface Properties on Adhesion. Applications to Medical Device Design

Polina Prokopovich[1,2]

Abstract Prediction of adhesion is of great significance in the development of micro-electromechanical systems and medical devices to achieve reliable and cost-effective design. For this purpose, knowledge of material and surface properties and their role on adhesion is crucial. This paper employs a multi-asperity adhesion model providing a greater understanding factors influencing on phenomena of adhesion and this novel method can be used as a tool for effective design of materials and their contact in various devices.

A dimensionless analysis, employing the π theorem, is presented based on the multi-asperity JKR adhesion model. The role of surface topography, material properties and the effect of asperity height distribution and its asymmetry on force of adhesion has been shown using dimensionless parameters. The application of the developed methodology is demonstrated through a case study on catheter design.

Introduction

Adhesion is a phenomenon that widely occurs in microelectronic and magnetic recording devices and emerging technologies such as optical data transmission switches used in microelectromechanical systems. Moreover, wear and frictional performance of materials in medical devices and orthopaedic implants are affected by adhesion [1–3]. It has been shown in previous works [4, 5], that force of adhesion is one of the most important factors affecting the lifetime of such devices. For example, stiction between contact surfaces leads to a major failure mode in MEMS as it breaks the actuation function of the device switches [6]. Hence, the understanding of adhesion is an essential requirement in designing and optimizing these micro-electromechanical systems [7]. In order to facilitate and predict materials properties and their design a significant knowledge of surface and material properties and their effect on adhesion is needed.

Traditionally, adhesive contact between two surfaces has been modelled by Johnson, Kendall and Roberts (JKR theory) [8] or by Derjaguin, Muller and Toporov (DMT theory) [9];. The JKR model has been the first one to account for surface energies of contacting solids whilst the DMT model considered long range adhesive forces outside the contact area. The latest theory is applicable for hard materials with low surface energy and small tip radii, whilst the JKR model is suited for soft materials with high surface energy and large tip radii. Maugis [10, 11] showed the continuous transition regime between the JKR and the DMT limits and provides a transitional solution for the intermediate cases between the JKR and DMT regimes. However, these models were derived for molecular smooth elastic solids. In practice, such ideal smooth surfaces do not exist due to manufacturing processes and, consequently, all surfaces possess morphological irregularities so called surface roughness. Due to the presence of surface roughness, a large number of asperity contacts should be considered [12–18].

The first model developed, that described the contact between two rough surfaces, was proposed by Zhuravlev in 1940 [19] and used for pure elastic contact; later Greenwood and Williamson (GW) applied the model to elasto-plastic contacts [15]. This model assumed that asperities have the same curvature radii and asperity heights described by a Gaussian distribution. These approximations proved to be not always correct, as it was shown in [20–24] that, some machining processes produce non-Gaussian surfaces with an asymmetric asperity heights distribution. Recently, Prokopovich and Perni [25] developed a multi-asperity adhesion model for two rough surfaces in contact based on the JKR- and DMT theories for a single asperity. In this later approach different roughness features, such as: asperity heights and

P. Prokopovich (✉)
[1]Welsh School of Pharmacy, Cardiff University, Redwood Building, King Edward VII Avenue, Cardiff, CF10 3NB, UK
[2]Institute of Medical Engineering and Medical Physics, School of Engineering, Cardiff University, Cardiff, UK
e-mail: prokopovichp@cf.ac.uk

curvature radii distributions, were considered and the model allows to predict adhesion between materials with any distribution of asperity heights and curvature radii. The validity of the model was verified against AFM force-distance measurements for several different materials, such as "soft"- PVC, silicone and "hard"- stainless steel and glass [25, 26]. Furthermore, a comparison of predictions between the DMT multi-asperity adhesion model and the corresponding JKR was made [26]. The later model differs from a model proposed by Borodich and Galanov [27] as the mechanical properties of the materials are required to predict the adhesive force, whilst in the Borodich and Galanov model, the adhesive and mechanical properties of the materials are estimated from experimentally obtained force–displacement curves.

In this work, the multi-asperity adhesion model developed by Prokopovich and Perni [25, 26], coupled with the π theorem has been utilised to determine the profile of contact force between two ideal solids with various material and surface properties such as roughness (asperity heights and its curvature radii), elasticity and surface energy. The effects of material and surface properties on adhesion have been studied. The dimensional multi-asperity contact analysis is applied to derive nondimensional relationships between adhesion and surface topography parameters and materials properties. In addition, variations of asperity heights and effect of their distributions on adhesion have been considered. The relevance of the proposed methodology is discussed using catheters as case study. The results are in good agreement with experimental measurements, which confirms the effectiveness of the model. The aim of this work is to propose simple rules for design purposes.

Modelling Approach

Based on this multi-asperity adhesion model [25, 26] the following nondimensional analysis was employed:

The numerical methodology is described as following:

- Ten thousand asperities were generated according to the cumulative distribution previously determined.
- Contact forces at separation distances (d), in the range 0–25 μm (dmax) in steps of 0.02 μm, between the two surfaces were estimated.
- Each asperity was analyzed for every distance and if its height is greater than d-δc, then the deformation of this asperity was determined and the contact force was estimated with the multi-asperity JKR adhesion model.
- The contact force of each contributing asperity was summed to determine the total adhesive contact force.

Determination of the Influencing Dimensionless Parameters

In this section the selection of the various dimensionless parameters required to describe a problem of adhesion is discussed. The π theorem [28] and the approach described in [25, 29] are utilised. In brief, the relevant definitions of the theorem are summarised below:

- Dimensional variables are the basic output of the test and they vary during an experiment.
- Dimensional parameters affect the variables and may differ from case to case, but they remain constant during a given run.

Based on the multi-asperity adhesion model described in Appendix 1 the following seven variables, whose unit in brackets, were derived:

$F(h)$ is the contact force [$N=kg\ m/s^2$]; Var
d is the separation distance [m]; Var
σ is the RMS surface roughness [m]; Par
μ is the average asperity height [m]; Par
R is the asperity curvature radius [m]; Par
$\Delta\gamma$ is the total surface energy [$J/m^2=N/m=kg/s^2$]; Par
E is the reduced elastic modulus [$Pa=N/m^2=kg/m\ s^2$]. Par

From all these set of variables there were only three different units defined (kg, m, s). Consequently, applying the π theorem, the number of the dimensionless parameters necessary to describe the problem was four (7−3) and the subsequent chosen dimensionless parameters were:

$$\underset{\text{Var}}{\frac{F(h)}{ER^2}},\quad \underset{\text{Par}}{\frac{d-\mu}{\sigma}},\quad \underset{\text{Par}}{\frac{\sigma}{R}},\quad \underset{\text{Par}}{\frac{E\sigma}{\Delta\gamma}}$$

Results and Discussions

Effect of Surface Topography

The influence of the surface characteristics on adhesion has been considered by varying the ratio of σ/R. The simulations were carried at a constant value of $\frac{E\sigma}{\Delta\gamma}$ equal to 12.5; this is in the range of the values obtained on the contact of biological tissues with catheter materials. A reduced elastic modulus (E) equal to 5×10^4 Pa and a total surface energy ($\Delta\gamma$) of 0.05 mJ/m^2 were chosen for the studied counter surfaces. In the simulation the following surface parameters, RMS surface roughness (σ) of 12.5 μm and average asperity height

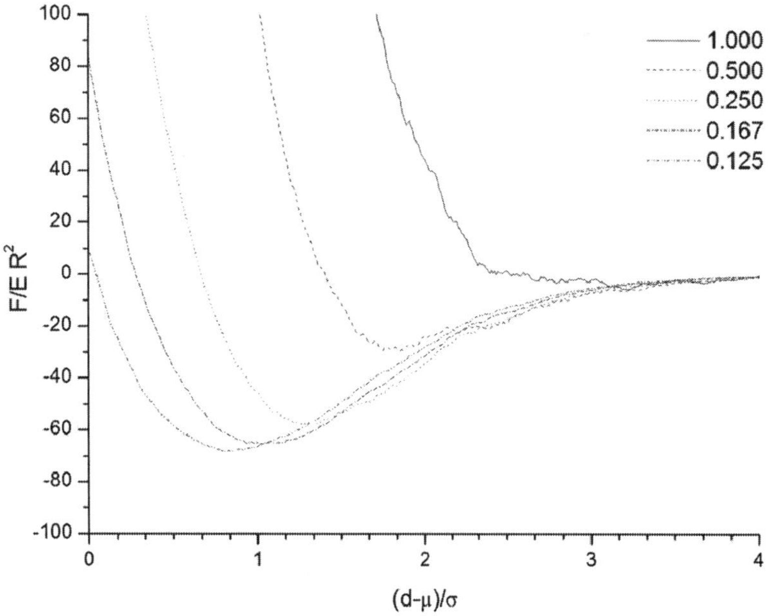

Fig. 1 Contact force against separation distance at various rations of σ/R for a value of $\frac{E\sigma}{\Delta\gamma} = 12.5$

(μ) of 25 μm were set constant, whilst the ratio of RMS roughness to asperity curvature radius was varied between 0.125 and 2.5. Results of dimensionless contact force $\left(\frac{F}{ER^2}\right)$ versus dimensionless separation distance $\left(\frac{d-\mu}{\sigma}\right)$ for different rations of σ/R are plotted in Fig. 1. This figure indicates that surface parameters have a significant effect on adhesion; with increasing ratio σ/R the contact force monotonically decreases.

Effect of Material Properties

The effect of the materials properties on the adhesion force has been taking into account through changing the ratio of $\left(\frac{E\sigma}{\Delta\gamma}\right)$. In this study the total surface energy was equal to 0.05 mJ/m^2 and the following surface features parameters were selected: σ of 12.5 μm; μ of 25 μm and asperity curvature radius(R) of 10 μm. Material elastic properties were varied though a reduced elastic modulus ranging from 2.0×10^4 Pa to 1.5×10^5 Pa. The variation of adhesion force for different ratios of $\frac{E\sigma}{\Delta\gamma}$ is shown in Fig. 2. With increasing $\frac{E\sigma}{\Delta\gamma}$ the force of adhesion decreased.

Elasticity and Plasticity Index

The pull-off forces derived from the contact force profiles at different values of σ/R are plotted against the elastic adhesion index(θ) are shown in Fig. 3a.

$$\theta = \frac{4}{3}\frac{E\sigma^{\frac{3}{2}}R^{0.5}}{R\Delta\gamma} \quad (1)$$

The change of the pull-off force becomes negligible above a certain value of θ. It has been reported a critical value θ of 10 which indicated a change of slope and a low adhesion beyond this value. However, other authors [31, 32] found that this critical value θ is not unique and depends on some parameters such as skewness values [32]. In this work the relation between elasticity adhesion index and the pull-off force followed the same pattern as in other works [31, 32] and the critical value of the elasticity adhesion index was about 20.

Similarly, in Fig. 3b, pull-off forces are plotted as function of plasticity index (ψ) Where H is the material hardness

$$\psi = \frac{E}{H}\left(\frac{\sigma}{R}\right)^{0.5} \quad (2)$$

The relation of the pull-off force against the plasticity index ψ is the same as in Fig. 3a; initially the pull-off force decreases with increasing plasticity index, but once a critical value is reached, the pull-off force remains almost constant with increasing values of the plasticity index. The critical value appeared to be about 30,000 times the value of surface hardness. Other authors [31, 32] reported different behaviours in the relation between plasticity index and pull-off force but this can be attributed to the choice of keeping the elasticity index constant made by those authors, whilst in this work, the elasticity index is left varying. Otherwise the simultaneous variation of other parameters in the simulation would be required.

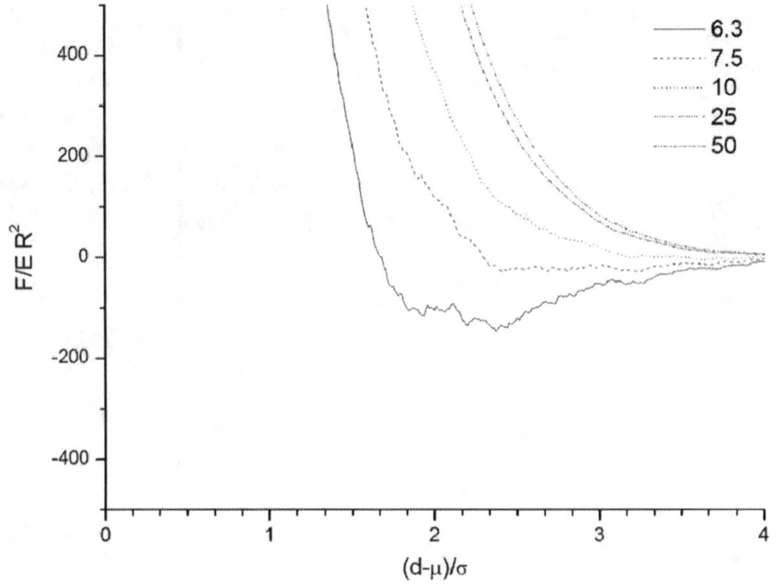

Fig. 2 Contact force against separation distance at various rations of $\frac{E\sigma}{\Delta\gamma}$ for a value of $\sigma/R = 2.5$

Effect of Asperity Height Distribution

It should be noticed that when non Gaussian distributions are used, the values of average and standard deviation do not possess any physical meaning, as opposed to normal distributions, though mathematically defined and possible to calculate. It is, therefore, advisable to ignore parameters such as σ or μ and use, instead, the parameters linked to the distribution chosen; for example, for a Weibull model, the shape parameter (p) and the size parameter (λ) should be employed (Eq. 3)

$$\phi(z) = \frac{p}{\lambda}\left(\frac{z}{\lambda}\right)^{p-1} \exp\left(-\frac{z}{\lambda}\right)^{p} \tag{3}$$

In this work the value of λ has been used in place of σ in the calculation of the non-dimensional pull-off force, whilst the nondimensional distance has been calculated as: $\frac{(d-\lambda)}{\lambda}p$.

The profile of contact force against nondimensional distance for a constant value of λ of 50 μm is shown in Fig. 4. It is evident that the values of pull-off force are quite affected by the variation of the parameter p; the force of adhesion increases with decreasing p and the separation distance resulting in the pull-off force increases with increasing p. This is cause by the different profile of the Weibull distribution with different p at constant λ; low values of p result in distributions with long tails, meaning that a significant number of asperities have great height. When the asperity heights are widely spread, it is more likely that only few asperities are deformed at the point corresponding to the pull-off force and its amount smaller than in case of asperity closely distributed. Each asperity contribution is individual hence, when the lowest asperity are engaged and giving negative contribution to the overall pull-off force, the highest asperity for are deformed to such a degree that their contribution can be positive to an extent offsetting the negative contributions, resulting in a positive contact force. This behaviour is the same as in case of the parameter σ/R.

Another way of characterising non Gaussian distribution is also through the Skewness (*Sk*). This parameter is used to measure the symmetry of the surface profile about the mean line. This number is sensitive to occasional deep valleys or high peaks. A symmetrical asperity height distribution has zero *Sk*. Surface roughness profiles with asperities removed or scrunched have a negative value of *Sk*, whilst profiles with high asperity heights or filled valleys have positive *Sk*. The values of *Sk* for the used distribution are shown in Table 1. Our results confirm that surfaces with positive *Sk* number exhibit low adhesion compared to surfaces with negative skewness numbers [33].

Case Study

The adhesion in catheters (silicone, polyurethane and PVC) will be analysed as example. Figure 3 demonstrated that the adhesion force is constant for values of elasticity index greater than 20. This value can be taken as the threshold for the catheter material selection.

$$\frac{E\sigma^{\frac{3}{2}}R^{0.5}}{R\Delta\gamma} > 20\frac{3}{4} = 15 \tag{4}$$

The value of Young modulus for aorta is: 100 KPa [2], whilst the surface energy for this blood vessel tissue is: 28.93 mJ/m^2 [2].

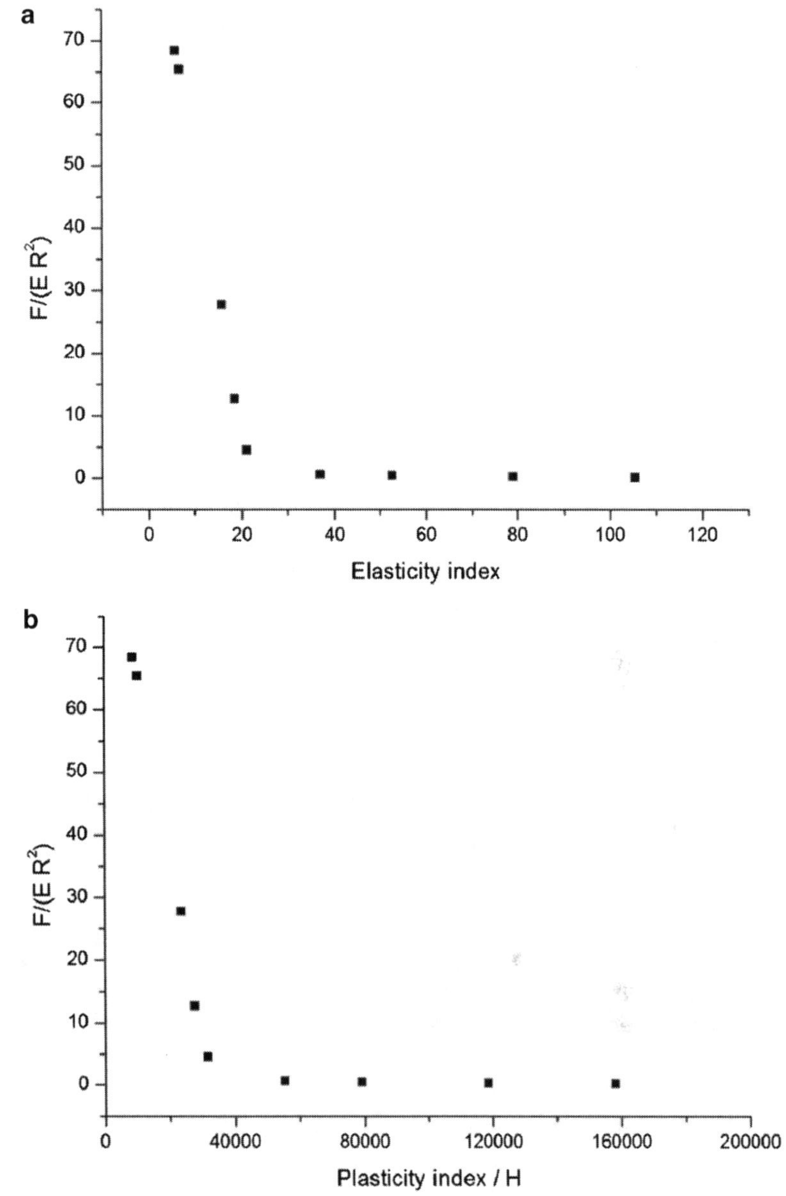

Fig. 3 Adhesion force as function of elasticity index (**a**) and plasticity index (**b**)

The composite Young modulus (E) is calculated as:

$$E = \left(\frac{1 - \upsilon_1^2}{E_1} + \frac{1 - \upsilon_2^2}{E_2} \right)^{-1} \tag{5}$$

Its value is practically corresponds to the Young modulus of aorta, because the Young's modulus of the elastomers used for catheters is few order of magnitude greater; for example, silicone is 0.5 MPa [2] and polyurethane 8 KPa [2]. The value of $\Delta\gamma$ is the sum of the surface energy of aorta and the catheter material; the later vary from 32 mJ/m^2 for silicone to 22 mJ/m^2 for polyurethane [2, 34], thus $\Delta\gamma$ remain almost unchanged among the elastomers.

Equation 4 can be rewritten as:

$$\frac{\sigma^{\frac{3}{2}}}{R^{0.5}} > 15 \frac{\Delta\gamma}{E} \tag{6}$$

this constitutes the designing equation for the surface topography of the catheter material as, using the assumption of Greenwood and Tripp [18]:

$$\sigma = \sqrt{\sigma_{material}^2 + \sigma_{aorta}^2} \tag{7}$$

And

$$\frac{1}{R} = \frac{1}{R_{material}} + \frac{1}{R_{aorta}} \tag{8}$$

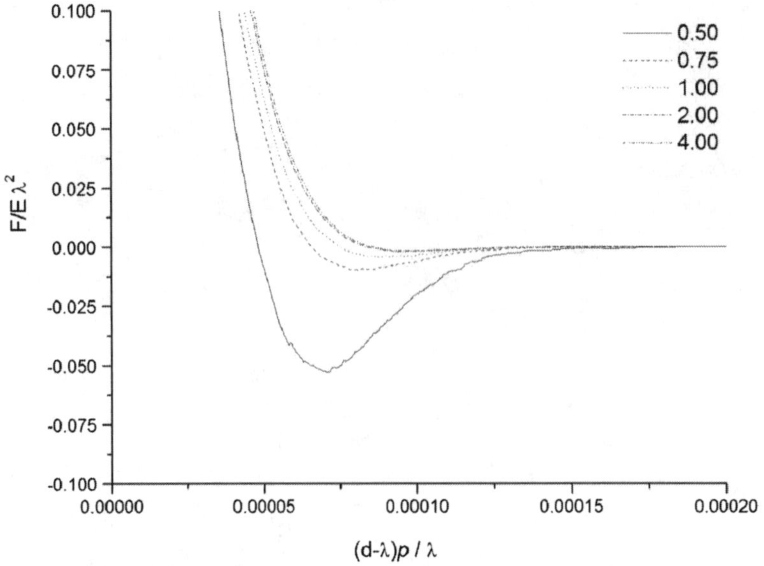

Fig. 4 Effect of shape parameter, p, of the asperity height distribution described by Weibull model on the contact force versus separation distance relation for λ=50 μm and $\frac{E\sigma}{\Delta\gamma} = 25$

Table 1 Skewness values for λ=50 μm with varying p

p	Sk
0.5	6.6
0.75	3.1
1.0	2.0
2.0	0.6
4.0	−0.09

Incorporating (7) and (8) into (6), the following relation is obtained:

$$\left(\sqrt{\sigma^2_{material} + \sigma^2_{aorta}}\right)^{\frac{3}{2}} \left(\frac{1}{R_{material}} + \frac{1}{R_{aorta}}\right)^{0.5} > 15\frac{\Delta\gamma}{E} \quad (9)$$

Data for the asperities properties of aorta can be taken from [2] who reported that R_{aorta} is 17.3 μm and σ is 3 μm. When the actual values of the known parameters are included in (8) the following relation is obtained (the materials properties are expressed in μm):

$$\left(\sqrt{\sigma^2_{material} + 9}\right)^{\frac{3}{2}} \left(\frac{1}{R_{material}} + 0.058\right)^{0.5} > 9 \quad (10)$$

Equation 10 describes a relation between surface roughness and asperity curvature radius in order to achieve the lowest force of adhesion between tissue and catheter. This relation can be employed in the design of catheters to minimise the undesirable effects caused by adhesion.

Conclusion

The results shown here extend the knowledge of how material and surface parameters influence adhesion using a multi-asperity adhesion model through unitless coefficients. The results are normalised using the elasticity and the plasticity index. For situations requiring low adhesion, the surface and material properties may be selected thus to yield high θ, high positive skewness and high ψ value for asymmetric rough surfaces.

Appendix 1 Brief Description of the Multi-Asperity JKR Adhesion Model

The JKR model [8] contains the deformation contribution resulting from the Hertz theory and an adhesion component due to surface energy $\Delta\gamma$. The multi-asperity adhesion model based on the JKR theory for a single asperity was developed by assuming that the situation could be simplied as a rough elastic surface in contact with a rigid smooth solid. This assumption has been validated in previous works by [17, 18]. The contact between these surfaces is considered as the sum of all individual asperity contacts [18]. Therefore, some asperities are deformed, whilst others are unaffected due to their small size and some asperities are stretched from their un-deformed height [25].

For a single asperity in contact with a flat surface the JKR theory predicts a contact force:

$$F_{JKR} = \frac{4Ea^3}{3R} - \sqrt{8\pi a^3 \Delta\gamma E} \tag{11}$$

The corresponding deformation is:

$$\delta_{JKR} = \frac{a^2}{R} - \frac{2}{3}\sqrt{\frac{\pi a \Delta\gamma}{E}} \tag{12}$$

The couple of (10) and (11) represent the relation, $F = fn(\delta, R)$ for the JKR model.

The Gaussian and Weibull distributions were employed to describe the population of asperity heights.

The overall contact force for a rough surfaces at a distance of d from a perfectly flat one is the sum of all the forces generated by the asperities whose height (h_i) are within their respective critical deformation $\delta_{c,i}$ [25]:

$$F(h) = \sum_{i}^{\delta_i > -\delta_{c,i}} fn(\delta_i, R_i) \tag{13}$$

where:

$$\delta_i = h_i - d \tag{14}$$

$$\delta_c = \frac{1}{3R}\left(\frac{9RF_c}{4E}\right)^{2/3} \tag{15}$$

$fn(\delta_i, R_i)$ is the adhesion force for one asperity with curvature radius R_i and deformation δ_i based on the JKR (Eqs. 10 and 11)

References

1. Prokopovich P, Theodossiades S, Rahnejat H, Hodson D (2010) Wear 268:845–852
2. Prokopovich P, Perni S (2010) Acta Biomat 6:4052–4059
3. Prokopovich P, Starov V (2011) Advances in colloid and interface science. 168(1–2):210–222
4. Maboudian R, Howe RT (1957) J Vac Sci Technol B 15:1–20
5. Bhushan B, Nosonovsky M (2004) Acta Mater 52:2461–2472
6. Wu L, Rochus V, Noels L, Golinval JC (2009) J Appl Phys 106:113502-1–113502-10
7. Scheu C, Gao M, Oh SH, Dehm G, Klein S, Tomsia AP, Rühle M (2006) J Mater Sci 41:5161–5168
8. Johnson KL, Kendall K, Roberts AD (1971) Proc Roy Soc A 324 (1558):301–313
9. Derjaguin KL, Muller VM, Toporov YP (1975) J Colloid Interface Sci 53(2):314–326
10. Maugis D (1992) J Colloid Interface Sci 150:243–269
11. Maugis D (1999) Contact, adhesion and rupture of elastic solids. Springer, New York
12. Cooper K, Ohler N, Gupta A, Beaudoin S (2000) J Colloid Interface Sci 222:63–74
13. Cooper K, Gupta A, Beaudoin S (2001) J Colloid Interface Sci 234:284–292
14. Eichenlaub S, Gelb A, Beaudoin S (2004) J Colloid Interface Sci 280:289–298
15. Greenwood JA, Williamson JBP (1966) Proc Roy Soc Lond A 295:300–319
16. McCool JI (1986) Wear 107:37–60
17. Francis HA (1977) Wear 45:221–269
18. Greenwood JA, Tripp JH (1971) Proc Inst Mech Eng 185:625–633
19. Zhuravlev VA (1940) Zh Tekh Fiz 10(17):1447–1452
20. Whitehouse DJ (1994) Handbook of surface metrology. Institute of Physics, Bristol
21. Yu N, Polycarpou AA (2002) ASME J Tribol 124:367–376
22. Yu N, Polycarpou AA (2004) ASME J Tribol 126:225–232
23. Stout KJ, Davis EJ, Sullivan PJ (1990) Atlas of machined surfaces. Chapman and Hall, London
24. Bhushan B (1999) Handbook of micro/nanotribology. CRC Press, Boca Raton
25. Prokopovich P, Perni S (2010) Langmuir 26(22):17028–17036
26. Prokopovich P, Perni S (2011) Colloid Surf A 383(1–3):95–101
27. Borodich FM, Galanov BA (2008) Proc Roy Soc A 464:2759–2776
28. Misic T, Najdanovic-Lukin M, Nesic L (2010) Eur J Phys 31: 893–906
29. White FM (1998) Fluid mechanics, 4th edn. McGraw Hill, New York
30. Sahoo P, Banerjee A (2005) J Phys D: Appl Phys 38:4096–4103
31. Patra S, Ali SM, Sahoo P (2008) Wear 265:554–559
32. Komvopoulos K (1996) Wear 200:305–327
33. Perni S, Prokopovich P, Piccirillo C, Pratten JR, Parkin IP, Wilson M (2009) J Mater Chem 19(17):2715–2723
34. Reedy ED, Starr MJ Jr, Jones RE, Flater EE, Carpick RW (2005) Proceedings of the 28th annual meeting adhesion society meeting, Mobile

Progr Colloid Polym Sci (2012) 139
DOI: 10.1007/978-3-642-28974-3_12

Influence of Anions of the Hofmeister Series on the Size of ZnS Nanoparticles Synthesised via Reverse Microemulsion Systems

Marina Rukhadze[1], Matthias Wotocek[2], Sylvia Kuhn[2], and Rolf Hempelmann[2]

Abstract Zinc sulfide nanocrystals with sizes of 4–7 nm were obtained by insufflation of hydrogen sulfide through reverse microemulsions, based on aqueous solutions of different zinc salts, nonionic surfactants and cyclohexane. The influence of the Hofmeister anions acetate, chloride, bromide, nitrate, iodide, and perchlorate on the micelles and thereof formed nanoparticles was studied by means of dynamic light scattering (DLS), X-ray diffractometry (XRD), UV-Vis spectroscopy and transmission electron microscopy (TEM). The sizes of micelles are significantly influenced by the kosmotropic or chaotropic nature of the actual anion, present in the water pools of reverse micelles: the diameter of the spherical ZnS nanoparticles, synthezised in these micelles, correlates with their size and thus follows the direction of the Hofmeister series. Several possible mechanisms are proposed to explain the influence of the anions.

Introduction

The synthesis of size controlled semiconductor nanoparticles, due to their unique optical, electronic and photocatalytic properties, is topical during the last two decades [1–9]. Among them, zinc sulfide nanoparticles have attracted some attention because of the potential application in photoelectronic transition devices [10]. The interest in this material brought about various preparation methods. In particular reverse microemulsions represent a well suited system for the synthesis of nanocrystalline materials owing to the facile control of size and shape, and, most notably, the narrow size distribution of the resulting nanoparticles [11–13].

R. Hempelmann (✉)
[1]Exact and Natural Sciences, I. Javakhishvili Tbilisi State University, Tbilisi 0128, Georgia
[2]Physical Chemistry, Saarland University, Saarbrücken 66123, Germany
e-mail: r.hempelmann@mx.uni-saarland.de

The Hofmeister series gives a rank order, which was established already in the nineteenth century. Originally it described the activity of different salt anions and cations to precipitate proteins [14] and was appointed as follows [15]:

$$SO_4^{2-} > F^- > OAc^- > Cl^- > Br^- > NO_3^- > I^- > ClO_4^- > SCN^- \quad (1)$$

This ranking which is also applicable to the solubility of hydrocarbons [16], the surface tension of water [17, 18] or the hydratisation of the ions [19], can be ascribed to the effect of these ions on the structure of water [19]. Kosmotropic ions, standing on the left side of the series, are small with a high charge density and bind nearby water molecules tightly, thus immobilizing them, what strengthens the structure of water built by hydrogen bonds and enhances hydrophobic effects. On the other hand chaotropic ions, standing on the right side, are large with low charge density and bind water only weakly, thus liberating nearby water molecules out of the water-structure by allowing more rapid motion than in bulk solution; that deranges the water-structure and lowers hydrophobic effects.

The structure of water is even more complex when it is confined to nanometre-scale cavities [20]. Water in the core of reversed micelles reveals at least two structures. Water that is close to the periphery of the micelle and thus in direct contact with the barrier molecules, i.e. surfactants, differs from water nearer to the centre of the reversed micelle. Thereby both of these structures differ from free, chemically pure water. Peripheral water molecules are more densely packed (high-density water) and form less hydrogen bonds when compared with free water. Water in the centre of the core is less densely packed (low-density water) and forms a larger number of hydrogen bonds than free water. The regular network of water molecules, which slightly resembles the molecular lattice of ice, is formed exactly by low-density water. The state of water inside the micelles may be considered as supercooled [21].

Although the Hofmeister series was discovered more than a century ago, it was neglected for a long time. The specific properties of ions were not considered in theories like, e.g., the DLVO-theory, where all ions of the background salt solution with the same ionic charge, due to the concept of ionic strength, should result in the same effective force between colloidal particles, regardless of the kind of the anions [22]. Only recently there is growing interest in these specific ionic effects, which are not completely understood until now [23].

The goal of the present work was to reveal the influence of anions of the Hofmeister series on the water structure in the water pools of the reversed micelles, which will be reflected in the sizes of both the micelles and the resulting nanoparticles of zinc sulfide, respectively.

Experimental

Materials

Nonionic surfactants (oxyethylene)$_4$ nonyl phenol ether (Tergitol NP-4, HLB~9, Fluka) and (oxyethylene)$_{10}$ nonyl phenol ether (Tergitol NP-10, HLB~13, Fluka), cyclohexane (VWR, p.a.), hydrogen sulfide (Fluka, 99%), zinc sulfate heptahydrate (Aldrich, 99%), zinc fluoride (Alfa Aesar, 97%), zinc acetate dihydrate (Fluka, 99%), zinc chloride (Riedel de Haen, 98%), zinc bromide (Merck, 98%), zinc nitrate hexahydrate (Grüssing, 99%), zinc iodide (Fluka, 98%), zinc perchlorate hexahydrate (ABCR, 99%), acetic acid (Aldrich, 25 wt.%, 99%), hydrofluoric acid (Riedel-de Haen, 40%, p.a.), hydroiodic acid (Merck, 57%), ethanol (Aldrich, 99%, 1% MEK), and deionised water. All chemicals were used as received without further purifications.

Preparation of Microemulsions

The preparation of the microemulsions is carried out in 50 mL measuring cylinders, standing in a water bath at 25°C. They are filled with 31 mL cyclohexane as the oil phase, 7.9 mL NP-5 and 3.9 mL NP-10 as surfactants, and 2, 3, or 4 mL aqueous solution of the respective zinc salt and then strongly agitated for several minutes. The obtained clear microemulsions are used the next day for synthesis. For the aqueous phase only freshly prepared solutions of zinc sulfate, zinc fluoride, zinc acetate, zinc chloride, zinc bromide, zinc nitrate, zinc iodide, or zinc perchlorate, respectively, with molarity 0.05 M are applied. Solutions of zinc fluoride, zinc acetate and zinc iodide become turbid because of hydrolysis, therefore they are acidified with hydrofluoric acid, acetic acid and hydroiodic acid in order to obtain a pH value of 5 like the other solutions, and in this way prevent precipitation.

Synthesis of ZnS Nanoparticles

For the synthesis, the respective microemulsions are transferred into a three-necked flask thermostated at 25°C. Under vigorous stirring H_2S is introduced into the solution by a capillary glass-tube for 45 min at an approximate rate of 2.5 bubbles per second in a bubble counter. Then ethanol is added to destroy the micelles and precipitate the formed nanoparticles, which are three times centrifuged at 3,400 rpm and washed with ethanol. In this step essentially all anions of the educts are removed in form of the corresponding acids. Finally the precipitates, i.e. the ZnS nanoparticles, are dried in vacuum at 30°C.

Characterization

The micelle sizes are determined by means of dynamic light scattering (DLS) at 25 °C at a scattering angle of 90° using an ALV-5000E set-up equipped with a frequency doubled neodymium-YAG laser (532 nm) purchased from Coherent Inc.. As samples the microemulsions have been taken as prepared. The refractive index of the hexane dispersion medium is 1.372, which is, concerning the scattering contrast, sufficiently different from the refractive index of the water in the micelles. For the DLS size determination the hexane viscosity of 0.294 mPas was used. In order to eliminate dust or other larger particles, the samples are filtered through syringe membrane filters with a pore size of 0.2 μm. TEM investigations are carried out on a JOEL JEM 2010 microscope working at 200 kV acceleration voltage; for that purpose the ZnS the powders are redispersed in cyclohexane and dried on a carbon-covered copper grid. X-ray diffractograms are measured by means of a Siemens D500 diffractometer using CuK_α radiation. For UV-Vis absorption the powders are suspended in cyclohexane (low solid content) and measured with a Perkin Elmer Lambda5 UV-Vis spectrometer.

Results and Discussion

By means of dynamic light scattering (DLS) the micelle sizes and their distribution are determined [24]. Thereby the field autocorrelation function is numerically evaluated

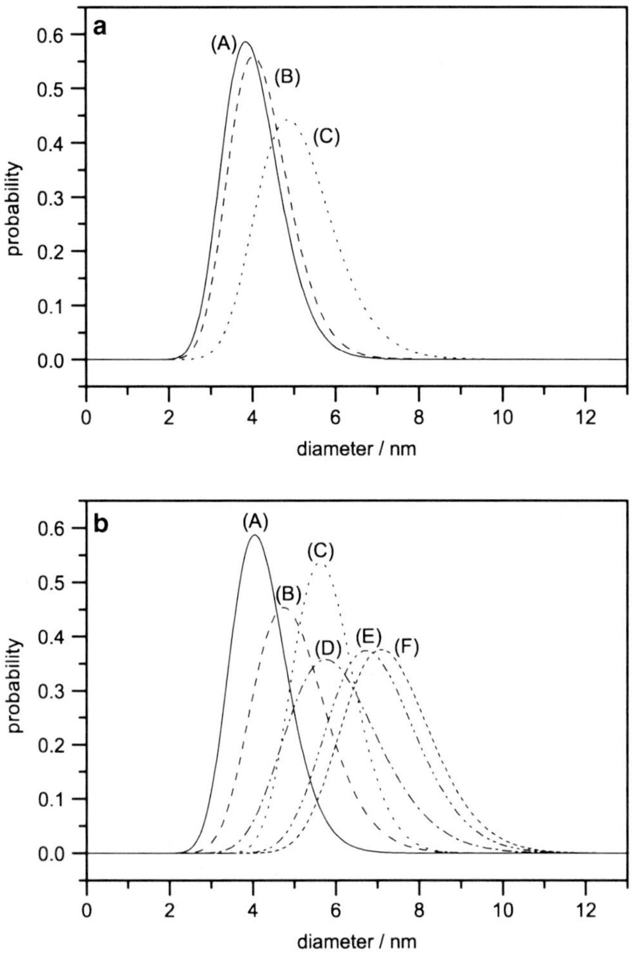

Fig. 1 (**a**) DLS results of micelle size distributions prepared with 0.05 M $Zn(OAc)_2$ solution: (A) 2 mL, (B) 3 mL and (C) 4 mL; and (**b**) DLS results of micelle size distributions prepared with 3 mL of 0.05 M solutions of (A) $Zn(OAc)_2$, (B) $ZnCl_2$, (C) $ZnBr_2$, (D) $Zn(NO_3)_2$, (E) ZnI_2 and (F) $Zn(ClO_4)_2$

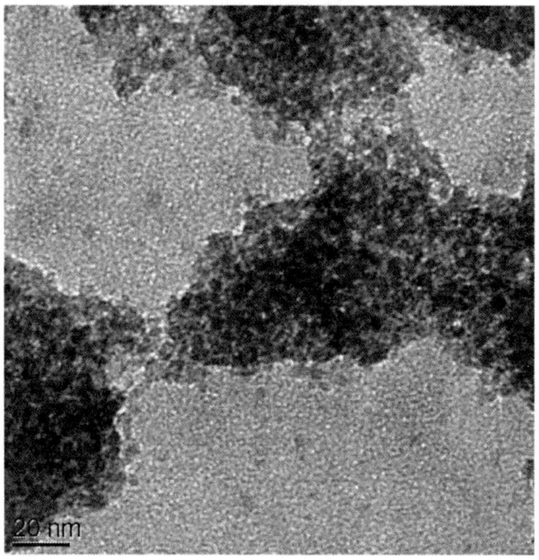

Fig. 2 Representative TEM image of the obtained ZnS nanoparticles (Prepared from $Zn(OAc)_2$)

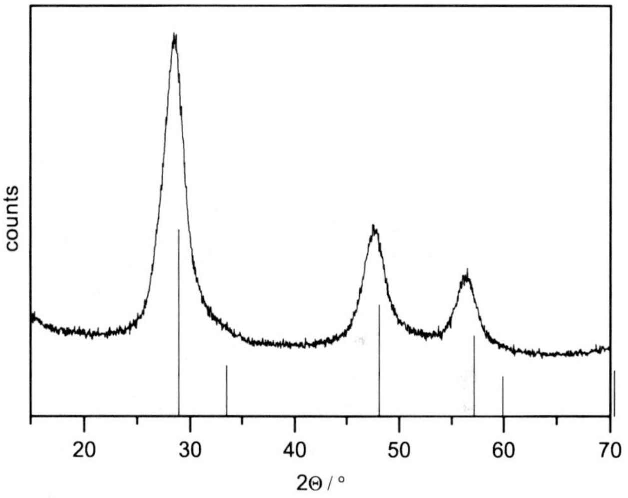

Fig. 3 X-ray diffractogram of synthesised ZnS nanoparticles

with the CONTIN algorithm [25], and for the purpose of comparison the results for the diameter distribution function are fitted (Fig. 1) using the log-normal distribution [26]:

$$g(D) = \frac{1}{\sqrt{2\pi}D \ln\sigma} \exp\left(-\frac{(\ln D - \ln\mu)^2}{2\ln^2\sigma}\right) \qquad (2)$$

Eventually, from the obtained median diameter μ and the relative width σ (geometrical width), the volume weighted average diameter is calculated (Fig. 6a) according to:

$$D_{vol} = \mu \ \exp\left(\frac{7}{2}\ln^2\sigma\right) \qquad (3)$$

As expected, raising the aqueous phase fraction of the microemulsions leads to an increase in the size of the micelles, because the given amount of surfactant can only stabilize a certain amount of surface area. This is exemplarily shown in Fig. 1a; similar features are observed for all microemulsions. Also the anion of the deployed zinc salt influences the micelle sizes for a given amount of water (Fig. 1b). Actually the sequence of this series corresponds exactly to the respective anions of the Hofmeister series.

The synthesized ZnS nanoparticles are then examined by transmission electron microscopy (TEM), and Fig. 2 shows one of the resulting micrographs which looks very similar for all zinc sulfide samples. Uniform nanospheres can be observed with a narrow size distribution, and the diameters are in good accordance to the sizes determined by XRD as outlined below. This agreement indicates that the ZnS nanoparticles are little singel crystals.

X-ray diffractograms (Fig. 3) of all ZnS nanoparticle samples are in good agreement with the available powder

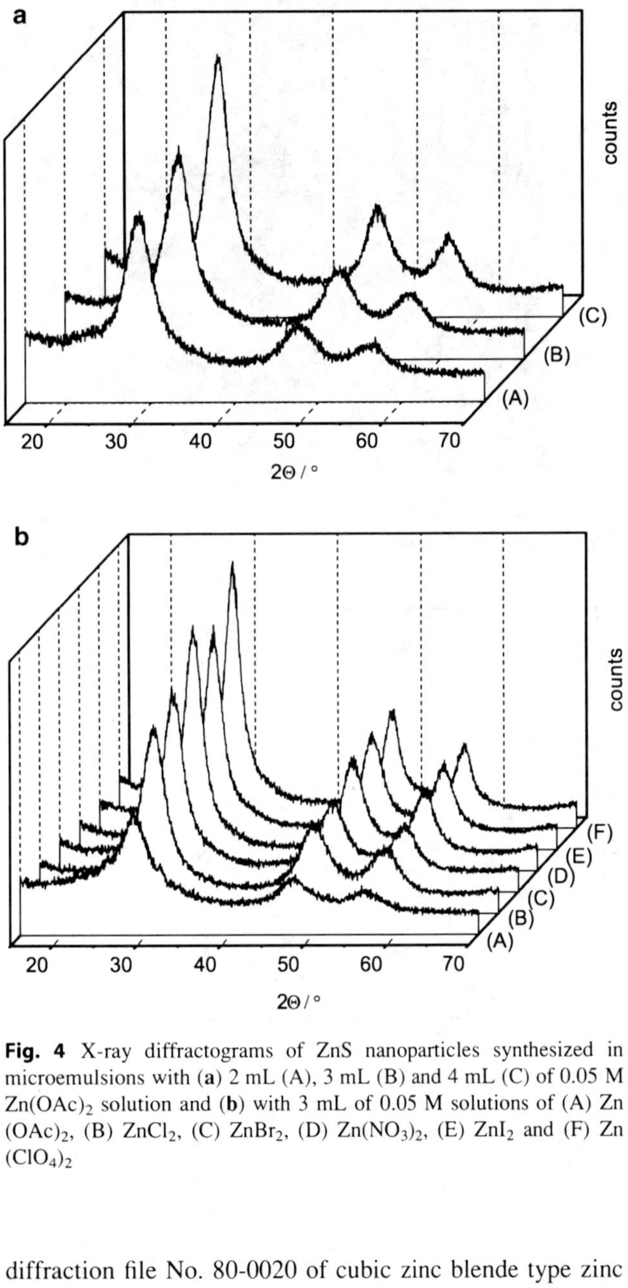

Fig. 4 X-ray diffractograms of ZnS nanoparticles synthesized in microemulsions with (**a**) 2 mL (A), 3 mL (B) and 4 mL (C) of 0.05 M $Zn(OAc)_2$ solution and (**b**) with 3 mL of 0.05 M solutions of (A) Zn $(OAc)_2$, (B) $ZnCl_2$, (C) $ZnBr_2$, (D) $Zn(NO_3)_2$, (E) ZnI_2 and (F) Zn $(ClO_4)_2$

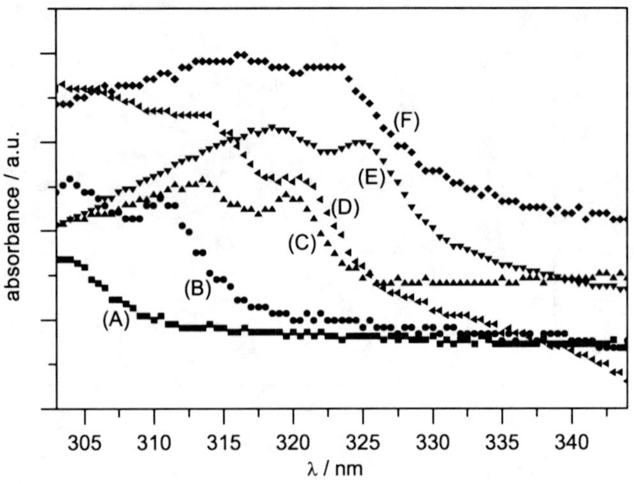

Fig. 5 UV-Vis-spectra of ZnS nanoparticles synthesized in microemulsions with 3 mL of 0,05 M solutions of (A) $Zn(OAc)_2$, (B) $ZnCl_2$, (C) $ZnBr_2$, (D) $Zn(NO_3)_2$, (E) ZnI_2 and (F) $Zn(ClO_4)_2$

diffraction file No. 80-0020 of cubic zinc blende type zinc sulfide with respect to the position and relative intensities of the peaks. The particles exhibit good crystallinity, and peaks corresponding to other crystalline material have not been found. The mean crystallite sizes are calculated (Fig. 6b) from the broadening of the (111) diffraction peak on the basis of the Scherrer equation [27] in the following way [26], assuming spherical particles:

$$D_{vol} = \frac{4}{3}\frac{K_{Scherrer}\cdot\lambda}{FWHM\cdot\cos\Theta} \tag{4}$$

Here D_{vol} is the volume-weighted mean diameter of the particles, K is Scherer's constant (set as K = 1), λ denotes the X-ray wavelength and FWHM is the full width at half maximum of the peak at the Bragg angle Θ, and the factor 4/3 transforms volume weighted column lengths into volume weighted diameters. Since the inverse micelles in the thermodynamically stable microemulsions represent nanoreactors and thus confine the particle growth, the particle sizes should correspond to the droplet sizes of the microemulsions determined above. In fact, with the increase of the amount of aqueous phase also the sizes of the particles increase, what is evident from the narrowing of the diffraction peaks in Fig. 4a. And furthermore the Hofmeister series is confirmed, too, in terms of the sequence of the particle sizes obtained from different zinc salts (anions) as educts (Fig. 4b).

In addition the semi-conductor quantum size effect is used to confirm the supposed Hofmeister effect by recording UV-Vis absorption spectra of the corresponding ZnS nanoparticles suspended in cyclohexane. Due to the poor dispersibility the particle content is very low and thus the spectra, unfortunately, are rather noisy. However, the effect can clearly be seen in Fig. 5: the absorption band is shifted to higher wavelengths in the Hofmeister sequence, what is caused by the larger particle sizes and therewith the reduction of the electronic band gap. Hence the radius r of the particles can be calculated approximately (Fig. 6b) using the concept of the "effective mass approximation" [2]:

$$E_g(r) = E_g(r\to\infty) + \frac{\hbar^2\pi^2}{2r^2}\left(\frac{1}{m_e^*}+\frac{1}{m_h^*}\right) - \frac{1.8\,e^2}{\varepsilon r} \tag{5}$$

For that purpose the respective measured band gap energy $E_g(r)$, the bulk band gap energy $E_g(r\to\infty)=3.62$ eV, the effective masses of the electron, $m_e^*=0.42\ m_e$, and the hole,

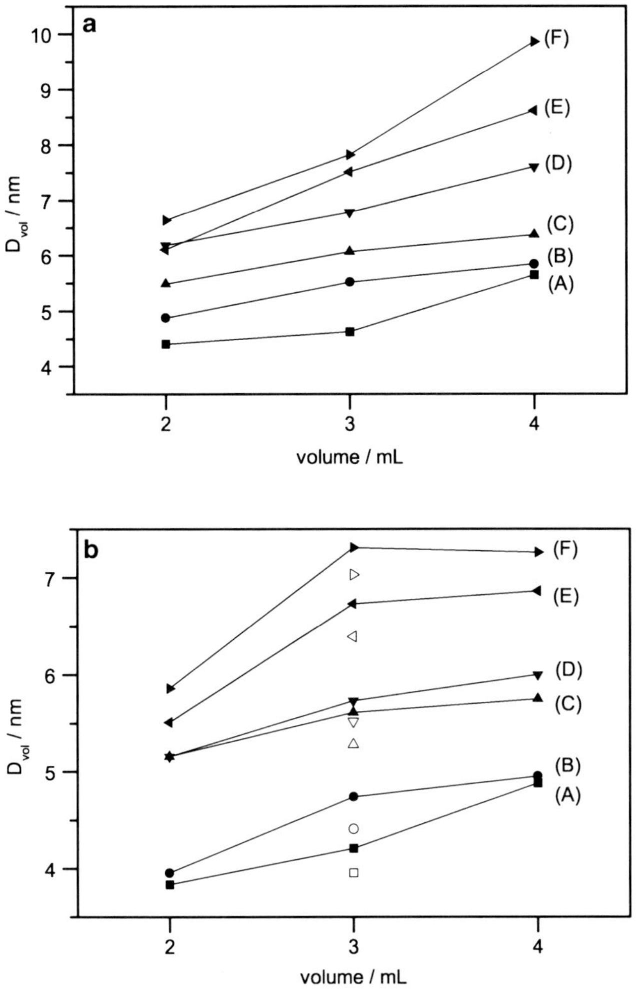

Fig. 6 (**a**) micelle sizes determined by means of DLS and (**b**) the corresponding ZnS crystallite sizes determined by XRD data (*full symbols*) and UV-Vis data (*empty symbols*) plotted versus the respective amount of 0.05 M solutions of (A) $Zn(OAc)_2$, (B) $ZnCl_2$, (C) $ZnBr_2$, (D) $Zn(NO_3)_2$, (E) ZnI_2 and (F) $Zn(ClO_4)_2$

$m_h^*=0.61\ m_e$, and the dielectric constant, $\varepsilon=5.20\ \varepsilon_0$, of ZnS are used [28].

In Fig. 6a and b all results are summarized; this diagram illustrates unambiguously that the micellar size and the resulting crystallite size of the zinc nanoparticles depend on the zink salt anions in accordance with the Hofmeister series. This effect may be explained by the different charge densities of the specific anions. Therefore chaotropic anions, which are only weakly hydrated and lower the hydrophobicity of water, may approach closely to the water/surfactant interface and thus penetrate even into the polar layer of surfactants. This leads to a rising of the intermolecular distances between the surfactants and results in the enlargement of the micelles. In contrast, the kosmotropic anions are hydrated abundantly and consequently migrate mostly in the higher-structured depth of the water pools, where they even enhance this order and thus the hydrophobic effect. Kosmotropic anions do not approach the surfactant layer, so the micelles are not expanded. The proposed considerations are supported by similar data on biological membranes [29]. Such comparison is possible, if one takes into account, that reverse micelles provide an artificial system that mimics the membranous biological system [30]. Molecular dynamics simulations were performed in order to reveal the Hofmeister series effect in the context of anion-lipid bilayer interactions. The central result from the simulations is that the large chaotropic anions penetrate more deeply into the interfacial region of the lipid bilayer interior since the larger ions are more hydrophobic and energetically stable in a hydrophobic environment; therefore they prefer the bilayer interior [29].

Conclusion

It was found that in reverse microemulsions both the amount of water and its structure are reflected in the sizes of the micelles and, subsequently, the sizes of nanoparticles synthesized in these micelles. The structure of water in the core of the reverse micelles is unambiguously determined by ionic additives, e.g. zinc salts with kosmotropic and chaotropic anions in the given case. Thus the results and the proposed interpretations may be useful for the study of ion-water interactions in confined spaces, viz. for biological objects, microemulsions, etc.

References

1. Efros AL, Efros AL (1982) Sov Phys Semicond USSR 16:772–775
2. Brus LE (1984) J Chem Phys 80:4403–4409
3. Henglein A (1989) Chem Rev 89:1861–1873
4. Horst W (1993) Angew Chem Int Edit Engl 32:41–53
5. Popović IG, Katsikas L, Weller H (1994) Polymer Bull 32:597–603
6. Alivisatos AP (1996) Science 271:933–937
7. Manzoor K, Vadera SR, Kumar N, Kutty TRN (2004) Appl Phys Lett 84:284–286
8. Michalet X, Pinaud FF, Bentolila LA, Tsay JM, Doose S, Li JJ, Sundaresan G, Wu AM, Gambhir SS, Weiss S (2005) Science 307:538–544
9. Rufino MNY, Galván MCÁ, Del Valle F, Villoria A, José M, José LGF (2009) ChemSusChem 2:471–485
10. Chen L, Shang Y, Xu J, Liu H, Hu Y (2006) J Dispers Sci Technol 27:839–842
11. Herrig H, Hempelmann R (1997) Nanostruct Mater 9:241–244
12. Xu C, Ni Y, Zhang Z, Ge X, Ye Q (2003) Mater Lett 57:3070–3076
13. Sottmann T, Strey R (2005) In: Lyklema J (ed) Fundamentals in interface and colloid science, vol 5. Elsevier, Amsterdam, ch. 5
14. Hofmeister F (1888) Arch Exp Pathol Pharmakol 24:247–260
15. Von Hippel PH, Wong KY (1964) Science 145:577–580
16. Long FA, McDevit WF (1952) Chem Rev 51:119–169

17. Weissenborn PK, Pugh RJ (1995) Langmuir 11:1422–1426
18. Baldwin RL (1996) Biophys J 71:2056–2063
19. Collins KD, Neilson GW, Enderby JE (2007) Biophys Chem 128:95–104
20. Levinger NE (2002) Science 298:1722–1723
21. Tsigankov VS, Sementin SA, Kucherenko AO, Okhotnikova LK (2002) Biofizika 47:863–865
22. Boström M, Deniz V, Franks GV, Ninham BW (2006) Advances in colloid and interface science 123–126:5–15
23. Hofmeister effects special issue (2004) Curr Opin Colloid Interf Sci 9:1–197
24. Beck Ch, Härtl W, Hempelmann R (1998) J Mater Res 13:3174–3180
25. Provencher SW (1982) Comput Phys Commun 27:213–227
26. Krill CE, Birringer R (1998) Philos Mag A 77:621–640
27. Scherrer P (1918) Göttinger Nachrichten 2:96–100
28. Gong S, Yao D, Jiang H, Xiao H (2008) Phys Lett A 372:3325–3332
29. Sachs JN, Woolf TB (2003) J Am Chem Soc 125:8742–8743
30. Chang GG, Hung TM, Hung HC (2000) Proc Natl Sci Counc Repub China B 24:89–100

Progr Colloid Polym Sci (2012) 139
DOI: 10.1007/978-3-642-28974-3_13

Polymer Shell Nanocapsules Containing a Natural Antimicrobial Oil for Footwear *Applications*

M.M. Sánchez Navarro, F. Payá Nohales, F. Arán Aís, and C. Orgilés Barceló

Abstract In this study, a series of melamine-formaldehyde (MF) nanocapsules containing essential oils as natural biocides with different polymer-to-oil ratio was prepared to be applied to footwear materials (lining, insoles, etc...) by an in situ polymerization (O/W) method. The nanocapsule physicochemical properties were characterized; the average size distribution was determined by DLS and the chemical structure was analyzed by FTIR spectroscopy. The antimicrobial effect of the essential oils was analyzed in solid media by measuring the inhibition diffusion halos in agar for 24 h. Finally, the incorporation of the synthesized nanocapsules into footwear materials was analyzed by scanning electron microscopy (SEM).

Introduction

Nanoencapsulation is a technique that has gained a strong foothold in different fields such as pharmaceuticals, cosmetics or food [1, 2]. Regarding the footwear sector, it is an emerging technology to be introduced in the near future, which will open new opportunities for the development of new functional materials with broad possibilities. Among its main advantages, this technique allows for extending the life of the active substances incorporated into the shoe, controlling the dose delivered over time (controlled release), etc. This will involve the production of footwear with improved properties for foot hygiene while serving as a differentiating element against other items.

Recently, interest in natural medicinal products, essential oils and other botanicals [3–5], has grown in response to the ever increasing incidence of adverse side effects associated with conventional drugs, and the emergence of resistance to antibiotics, synthetic disinfectants and germicides. Some essential oils such as tea tree, lemon, lavender, eucalyptus etc, show suitable antimicrobial activity to be used as biocide [3, 4] for different applications.

Furthermore, the in situ polymerization allows for the formation of nanocapsules containing water-immiscible dispersed phase, with improved mechanical properties [6] and thermal stability [7]. The properties of the membrane depend not only on its chemical structure but also on all the synthesis conditions. The polycondensation of the amino resin occurs in the continuous phase, and the polymer precipitation around the oil droplets to form nanocapsules shell is linked to the pH and the melamine-formaldehyde molar ratio [8, 9].

In this study, melamine-formaldehyde MF nanocapsules containing tea tree essential oil as natural biocide were prepared by in situ polymerization. This work aims to investigate the synthesis and properties of MF nanocapsules containing this essential oil as a function of their melamine-formaldehyde/oil ratio.

Experimental

Materials

Melamine and formaldehyde were obtained from Quimidroga S.A. (Barcelona, Spain) and used as received. Essential oils (EO) were purchased from Guinama S.L.U. (Alboraya, Valencia, Spain). Sodium dodecylsulphate (SDS) from Sigma-Aldrich was used without further purification. Water was purified with Millipore automatic Sanitization Module.

Melamine-Formaldehyde Resin Preparation

Prior to the encapsulation, 3 g of melamine and 30 mL of water were added in a round-bottom glass flask. The vessel

M.M.S. Navarro (✉)
INESCOP. Footwear Research Institute, Pol. Ind. Campo Alto s/n., Elda(Alicante) 03600, Spain
e-mail: msanchez@inescop.es

was immersed in a water bath placed on a magnetic stirrer with heating. When the temperature reached 50°C, 6 mL of formaldehyde were added and the mixture was magnetically stirred for 1 h at 70°C. Finally, a melamine-formaldehyde resin clear solution was obtained as prepolymer (MF).

Synthesis of Melamine-Formaldehyde Nanocapsules Containing Essential Oils

First of all, an O/W emulsion was prepared. The oil phase was composed of an essential oil and the water one was constituted of sodium dodecylsulphate (SDS) as surfactant and distilled water. Both phases were mixed and then emulsified by a sonifier (Branson) during 90 s at 50% amplitude. The result was a milky emulsion. Subsequently, the emulsion was maintained at 45°C and magnetically stirred, while the MF resin prepolymer obtained in the previous section was added drop by drop. After the addition, the pH of the mixture was adjusted to 4 and the emulsion was left stirring for 1 h. After this time, the temperature was risen to 85°C and kept stirring for 2 h. Finally, it was allowed to cool down to room temperature and the pH was adjusted to 10.

Finally, the synthesized nanocapsules were freeze-dried, obtaining a fine white powder for further characterization through different experimental techniques.

In this study, nanocapsules with different MF/oil ratios (1 to 7) were synthesized and characterized. The synthesized nanocapsules containing tea tree oil (EO) were referenced as: MF1EO, MF2EO, MF3EO, MF4EO, MF5EO, MF6EO and MF7EO with 1 to 7 MF/EO ratios respectively.

Nanocapsule Characterization

The particle size distribution and mean average particle size of the nanocapsules containing EO were determined using a laser particle analyzer Coulter LS 230 with a small volume module. FTIR spectroscopy was used to analyze chemical properties of the freeze-dried nanocapsules by transmission mode in KBr pellets. For the morphological characterization and evaluation of the incorporation into different footwear materials, scanning electron microscopy (SEM) was carried out using a Philips EM400 microscope operating at 80 kV. Previously, specimens were gold coated to obtain a good contrast.

Results and Discussion

Antimicrobial Activity Study

Prior to the encapsulation process, the antimicrobial activity of three different essential oils (almond, Tea Tree and lemon oils) against different microorganisms typically found in used footwear (*Escherichia coli* SG13009 (QIAGEN), *Bacillus subtilis* 168 ATCC 23857, *Klebsiella pneumoniae* CECT 141 and *Staphylococcus aureus* CECT 239) [10] was analyzed by in vitro assays. The evaluation was carried out by measuring their activity in liquid media and inhibition halos by agar diffusion for 24 h.

The antimicrobial activity was tested through measurement of the absorbance at 600 nm in liquid medium. Firstly, the different microorganisms were inoculated in LB medium (*Escherichia coli* and *Bacillus subtilis*) and nutrient broth (*Klebsiella pneumoniae* and *Staphylococcus aureus*), respectively. Afterwards, three tubes were inoculated for each species and the following substances were poured into them: almond oil as blank and two different essential oils (Tea Tree and lemon oil). Moreover, an oil-free tube was inoculated as negative control. All of them were grown up at 37°C and the absorbance at 600 nm after 24 h of growth was measured.

The evaluation of the inhibition halos assay was carried out by seeding four medium plates for each type of bacteria. A disc of cloth impregnated with the above-mentioned essential oils was placed on them, as well as a control one with no antimicrobial substance. After 24 h of growth, the inhibitory halos were measured. Table 1 shows the response of the microorganisms to the different oils in order to test the antimicrobial activity. The results proved the effective antimicrobial activity of Tea Tree oil against most of the

Table 1 Antimicrobial activity test results for different oils and for the control with no-antimicrobial substance

Sample	*E. coli*		*B. subtilis*		*K. pneumoniae*		*S. aureus*	
	Liq	Solid	Liq	Solid	Liq	Solid	Liq	Solid
Control	–	–	–	–	–	–	–	–
Almond oil	–	–	–	–	–	–	–	–
Lemon oil	+	–	++	–	+	++	++	+
Tea tree oil (TTO)	++	–	++	++	+++	–	+++	++

(–): no antimicrobial activity observed
(++): moderate antimicrobial activity observed
(+++): strong antimicrobial activity observed

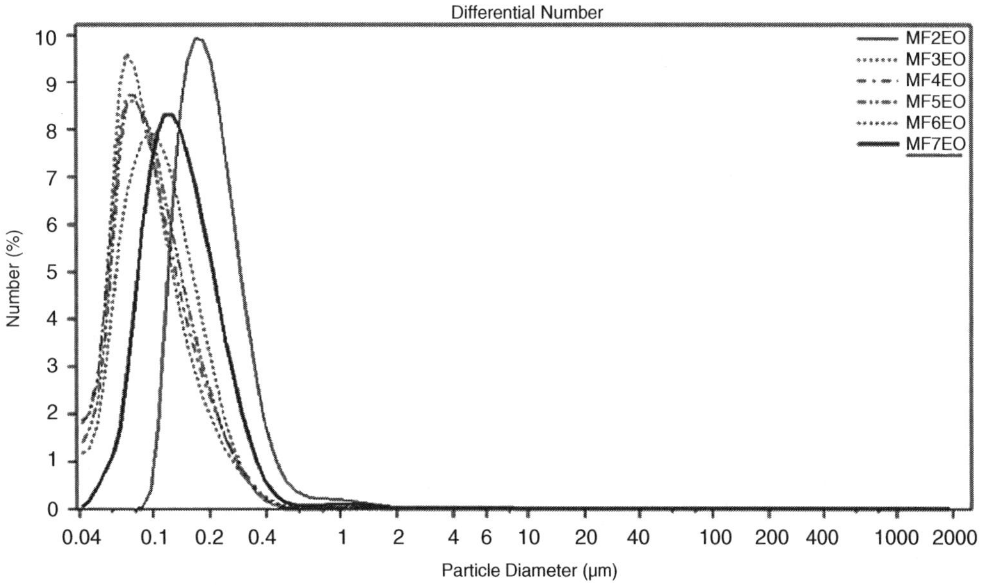

Fig. 1 Average particle size distribution of the EO nanocapsules with different MF/EO ratios

microorganisms. For that reason, this essential oil (EO) was chosen for encapsulation.

Nanocapsule Characterization

Figure 1 shows the average particle size distribution in number for nanocapsule emulsions synthesized using different MF/EO ratio. They showed a monomodal distribution with an average mean particle size around 400 nm but with a small amount of capsules greater than 1 µm. In general, mean particle size tends to increase as MF/ratio increases. Nevertheless, MF2EO showed a strange behaviour with a greater mean particle size, maybe due to a low efficiency of the nanoencapsulation process for low MF/EO ratios.

These nanoparticles show a mononuclear core-shell morphology (nanocapsules) observed by SEM of a freeze fracture of the solid particles in a thermoset resin matrix. After that, EO load in the nanocapsules was determined solvent extraction using dichloromethane. EO load (%) was calculated as [extracted EO from the nanocapsules]/[theorical EO in the nanocapsule] ratio. EO load increases as MF/EO ratio increases from 13.7 to 68.7 (weight %).

Figure 2 shows the FTIR spectra corresponding to MF resin shell, EO and MFEO nanocapsules with two different MF/EO ratios, 2 and 7.

For the MF resin sample, a characteristic broad band responsible for hydroxyl, imino and amino groups stretching was observed around 3335 cm^{-1}. Alkyl C-H stretching vibration was found around 2930 cm^{-1}. The C-N multiple stretching in the triazine ring was observed around 1537 cm^{-1}. C-H bending vibration in CH_2 was found at 1450 and 1369 cm^{-1} due to methylene bridges. The characteristic absorption bands of aliphatic C-N vibration appeared between 1200 and 1170 cm^{-1}. Characteristic triazine ring bending at 812 cm^{-1} could also be observed.

EO spectrum showed a broad band responsible of hydroxyl stretching around 3430 cm^{-1}. Alkyl C-H stretching vibration was found between 2960 and 2854 cm^{-1}. The band at 1744 cm^{-1} corresponds to C=O vibration bond. Between 1464 and 1443 cm^{-1} the C-H bending vibration was observed. Characteristic absorption band of vibration C-O-C aromatic rings appeared at 927 cm^{-1}.

FTIR spectra of the nanocapsules containing EO proved to be the physical combination of the characteristic bands of MF resin and the EO selected. As the resin ratio increases, the characteristic bands of the resin increase, and the oil bands decrease. This seems to be due to an increased thickness of the nanocapsule shell, according to previous results in the literature.

Finally, the nanocapsules containing the tea tree oil were incorporated into some materials as leather and fabrics commonly used in the footwear industry. The incorporation to the footwear materials was done by immersion of the materials in an aqueous dispersion containing the synthesized nanocapsules without binder. In that process, a piece of each material was immersed into 50 mL of the nanocapsules dispersion for 1 h under mechanical agitation at room temperature. After that, the samples were left to dry at room temperature overnight. The distribution and anchorage of the nanocapsules on material surface was analysed by SEM.

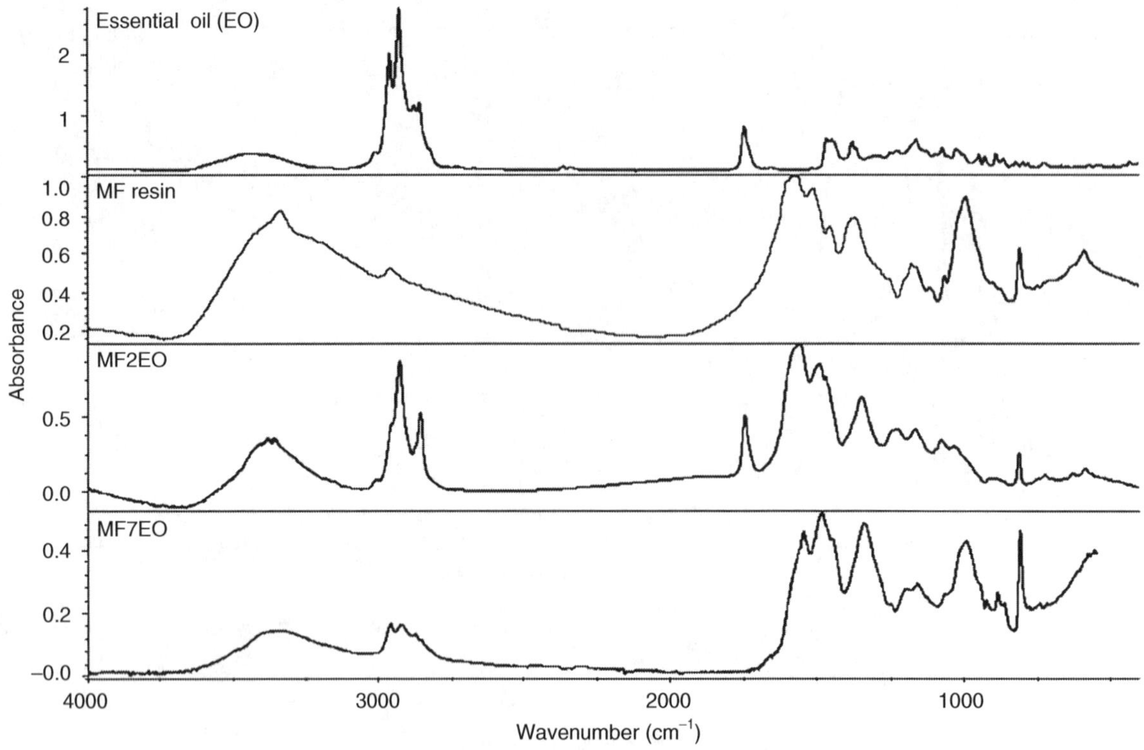

Fig. 2 FTIR spectra of the EO, MR resin and freeze-dried nanocapsules with different MF/EO ratio

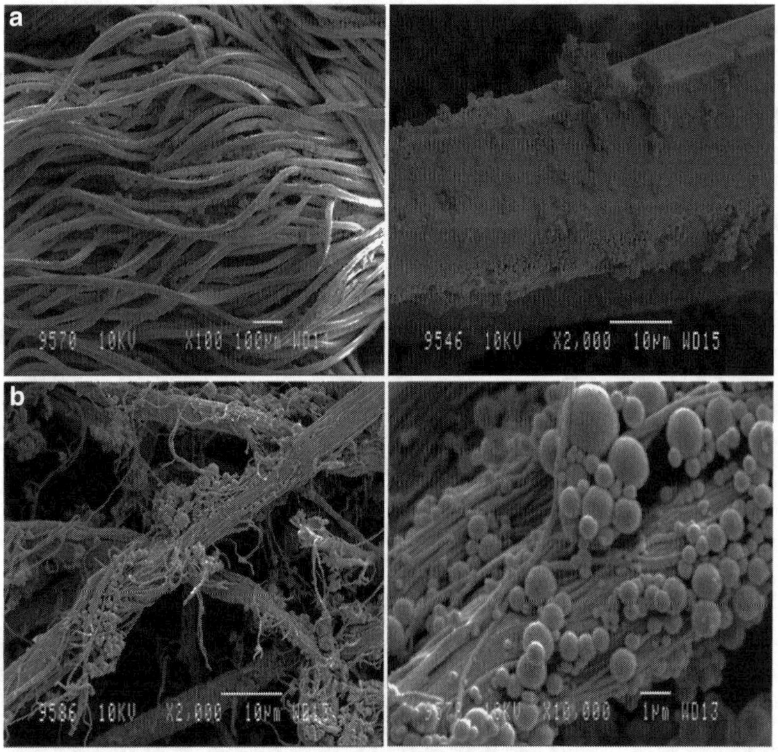

Fig. 3 (**a**) SEM images of the EO nanocapsules attached to the fabric fibers; (**b**) SEM images of the EO nanocapsules attached to the leather fibers

Figure 3a and b show SEM images of the nanocapsules attached to fabric fibres and leather, respectively. In further work, the use of conventional finishing techniques as well as a suitable binder is being investigated in order to improve anchoring of the nanocapsules to the footwear materials and therefore their durability.

Conclusions

Firstly, some essential oils showed suitable antimicrobial activity against typical microorganisms associated to the foot skin, which indicates that they could be used as natural antimicrobial agents for footwear application.

Secondly, the nanoencapsulation of an essential oil represents an innovative solution for the controlled release of this active substance in footwear. Nevertheless, further optimization of the characteristics would be needed in order to get the best properties for a long lasting antimicrobial effect in footwear applications, depending on the type of shoe, user and materials.

Acknowledgements The Research was partially funded by the Spanish Ministry of Science and Innovation (project nº CTQ 2010-16551).

References

1. Benita S (1996) Microencapsulation: methods and industrial applications. Marcel Dekker, New York, pp 1–3
2. Kumar Gosh S (2006) Functional coatings by polymer microencapsulation. Wiley–VCH Verlag GmbH & Co. KGaA, Weinheim, pp 12–15
3. Kunicka-Styczynscka A, Sikora M, Kalemba D (2009) Antimicrobial activity of lavender, tea tree and lemon oils in cosmetic preservative systems. J Appl Microbiol 107:1903–1911
4. Reuter J et al (2010) Botanicals in dermatology, an evidence-based review. Am J Clin Dermatol 11:247–267
5. Saleem M, Nazir M, Ali MS, Hussain H, Lee YS, Riaz N, Jabbar A (2010) Antimicrobial natural products: an update on future antibiotic drug candidates. Nat Prod Rep 27:238–254
6. Sun G, Zahng Z (2001) Mechanical properties of melamine-formaldehyde microcapsules. J Microencapsul 18:593–602
7. Luo W, Yang W, Jiang S, Feng J, Yang M (2007) Microencapsulation of decabromodiphenyl ether by in situ polymerization: preparation and characterization. Polym Degrad Stabil 92:1359–1364
8. Lee HY, Lee SJ, Cheong W, Kim JH (2002) Microencapsulation of fragrant oil via in situ polymerisation: effects of pH and melamine-formaldehyde molar ratio. J Microencapsul 19:559–569
9. Jun-Seok H (2006) Factors affecting the characteristics of melamine resin microcapsules containing fragrant oils. Biotechnol Bioprocess Eng 11:391–395
10. Cuesta Garrote N, Sánchez Navarro MM, Arán Aís F, Orgilés Barceló C (2010) Natural antimicrobial agents against the microbiota associated with insoles. Science and Technology Against Microbial Pathogens Research, Development and Evaluation. In: Mendez-Vilas A (ed) Proceedings of the international conference on antimicrobial research (ICAR2010), Valladolid, Spain, pp 109–113

Progr Colloid Polym Sci (2012) 139
DOI: 10.1007/978-3-642-28974-3_14

Evaporation of Pinned Sessile Microdroplets of Water: Computer Simulations

S. Semenov[1], V.M. Starov[1], R.G. Rubio[2], and M.G. Velarde[3]

Abstract The aim of present work is to describe results of computer simulations, which show the influence of kinetic effects on evaporation of pinned sessile submicron droplets of water. The suggested model takes into account both diffusive and kinetic regimes of evaporation. The obtained results show a smooth transition between kinetic and diffusive regimes of evaporation as the droplet size decreases from millimetre to micrometer size.

Introduction

The evaporation of sessile liquid droplets plays a significant role in practical applications such as spray cooling [1, 2], ink-jet printing [3], tissue engineering [4], printing of microelectromechanical systems (MEMS) [5], surface modification [6, 7], various coating processes [8], as well as biological applications [9]. It is a reason why a number of theoretical and experimental investigations have been focussed on this phenomenon [10–18].

Studying of evaporating of microdroplets can help to understand the influence of Derjaguin's (disjoining/conjoining) pressure acting in a vicinity of the apparent three-phase contact line [19–21].

The aim of present computer simulations is to show how the evaporation of pinned sessile submicron size droplets of water on a solid surface differs from the evaporation of bigger millimetre size droplets. The obtained results prove the importance of kinetic effects, whose influence becomes more pronounced for submicron droplets

V.M. Starov (✉)
[1]Department of Chemical Engineering, Loughborough University, Loughborough LE11 3TU, UK
e-mail: V.M.Starov@lboro.ac.uk
[2]Department of Química Física I, Universidad Complutense, Madrid 28040, Spain
[3]Instituto Pluridisciplinar, Universidad Complutense, Madrid 28040, Spain

The model used below includes both diffusive and kinetic regimes of evaporation. Our model differs from a purely diffusive model because Hertz-Knudsen-Langmuir equation [22, 23] is used for a boundary condition at the liquid–vapour interface instead of a saturated vapour at the liquid–vapour boundary. The adopted model differs from a purely kinetic model of evaporation because the model also includes the presence of vapour diffusion into the surrounding vapour similar to the diffusion model. In this way the evaporation rate is controlled by both rates of vapour diffusion into ambient vapour and molecules transfer across the liquid–vapour interface. As a result the vapour concentration at the liquid–vapour interface falls in between saturated value and the ambient value. The vapour concentration takes the value that drives both diffusive and kinetic local fluxes at the liquid–vapour interface. According to the mass conservation law these two fluxes are equal to each other.

Computer simulations are performed using the software COMSOL Multiphysics. The dependences of total molar evaporation flux, J_c, on the radius of the contact line, L, and the value of contact angle, θ, are studied below.

Problem Statement

Below only very small droplets are under consideration that is the influence of gravitation is neglected. The problem under consideration has an axial symmetry. It is assumed that sessile droplet forms a sharp three-phase contact line and maintains a spherical-cap shape of the liquid–vapour interface due to the action of liquid–vapour interfacial tension. A geometry of the problem is presented schematically in Fig. 1.

Due to small size of evaporating droplets under consideration (radius of the contact line, L, is less than 1 mm), the diffusion of vapour in the vapour phase dominates the convection. The latter is confirmed by small values of both thermal

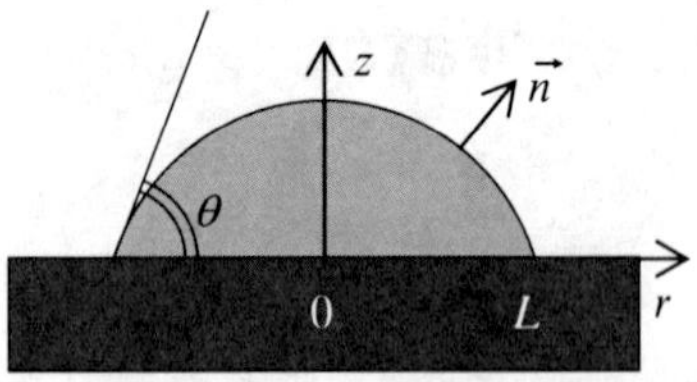

Fig. 1 An axisymmetric sessile droplet on a solid substrate. θ and L are contact angle and radius of the droplet base

and diffusive Peclet numbers: $\mathrm{Pe}_\kappa = Lu/\kappa < 0.05$; $\mathrm{Pe}_D = Lu/D < 0.04$, where L is the radius of the contact line, u (< 1 mm/s) is the characteristic velocity of vapour convection due to evaporation and Marangoni convection, κ is the thermal diffusivity of the surrounding air, and D is the diffusion coefficient of vapour in air at normal conditions.

Hence, the convective terms in transfer equations inside the vapour phase are neglected below. As in Ref. [24] the problem is solved under quasi-steady state approximation. That is, all time derivatives in all equations are neglected. The quasi-steady solution of the problem gives simultaneous distribution of heat and mass fluxes in the system.

A volume of the droplet decrease over time, which is caused by the evaporation. Therefore, a certain velocity is ascribed to the liquid–vapour interface in order to satisfy the mass conservation law in the quasi-steady state approximation. This velocity is calculated in such a way to preserve the spherical-cap shape of the liquid–vapour interface and to match the evaporation rate of the droplet.

The parameters of the following materials are used in present computer simulations: copper as the substrate, water as the liquid droplet, and a humid air as a surrounding medium. The pressure in the surrounding gas is equals to the atmospheric pressure, the ambient temperature is 20°C, and the ambient air humidity is 70%.

Governing Equations in the Bulk Phases

The following governing equations describe the heat and mass transfer in the bulk phases: equation of the heat transfer in solid and gas phases:

$$\Delta T = 0,$$

where Δ is the Laplace operator, and T is the temperature; equation of the heat transfer inside the liquid droplet:

$$\vec{u} \cdot \nabla T = \kappa_l \, \Delta T,$$

where $\vec{u}$ is the liquid velocity, ∇ is the gradient operator, and κ_l is the thermal diffusivity of the liquid; Navier-Strokes equations are used for a flow inside the liquid droplet:

$$\rho_l \, \vec{u} \cdot \nabla \vec{u} = \nabla \cdot \mathbf{T},$$

where ρ_l is the liquid density, $\nabla \vec{u}$ is the gradient of the velocity vector, $\mathbf{T}$ is the full stress tensor, and $\nabla \cdot \mathbf{T}$ is the dot-product of nabla operator and the full stress tensor. The full stress tensor is expressed via hydrodynamic pressure, p, and viscous stress tensor, $\boldsymbol{\pi}$ as:

$$\mathbf{T} = -p\mathbf{I} + \boldsymbol{\pi},$$

where $\mathbf{I}$ is the identity tensor. Continuity equation for the liquid phase is used also:

$$\nabla \cdot \vec{u} = 0.$$

Diffusion equation for the vapour in the gas phase:

$$\Delta c = 0, \tag{1}$$

where c is the molar concentration of the vapour.

Boundary Conditions

Boundary conditions of temperature continuity are applied at all interfaces (liquid–vapour, liquid–solid and gas–solid):

$$T_l = T_g, \quad T_l = T_s, \quad T_g = T_s,$$

where subscripts l, g and s stand for liquid, gas and solid, respectively. Continuity of the heat flux is applied on solid–liquid and solid–vapour interfaces:

$$-k_l(\nabla T)_l \cdot \vec{n} + k_s(\nabla T)_s \cdot \vec{n} = 0,$$

$$-k_g(\nabla T)_g \cdot \vec{n} + k_s(\nabla T)_s \cdot \vec{n} = 0,$$

where k is the thermal conductivity of corresponding phase; $\vec{n}$ is the unit vector, perpendicular to a corresponding interface. At the liquid–vapour interface heat flux experiences discontinuity caused by the latent heat of vaporization:

$$-k_l(\nabla T)_l \cdot \vec{n} + k_g(\nabla T)_g \cdot \vec{n} = j_c \Lambda,$$

where Λ is the latent heat of vaporization (or enthalpy of vaporization [25], units: J/mol); $\vec{n}$ is the unit vector, normal to the liquid–vapour interface, and pointing into the vapour phase; j_c is the surface density of the molar flux of evaporation ($\mathrm{mol \cdot s^{-1} \cdot m^{-2}}$) at the liquid–vapour interface.

No-slip and no-penetration boundary conditions are used for Navier–Stokes equations at the liquid–solid interface, resulting in zero liquid velocity at this interface:

$$\vec{u}_l = 0.$$

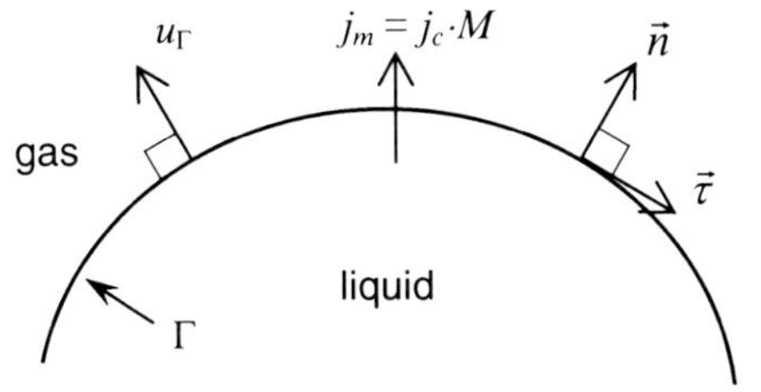

Fig. 2 Notations at the liquid–vapour interface. Γ is the liquid–vapour interface; u_Γ is the normal velocity of the interface Γ in direction of normal unit vector $\vec{n}$ (from the liquid phase to the gaseous one); $\vec{\tau}$ is the unit vector, tangent to the interface Γ; j_m is the mass flux across the interface Γ

Let Γ is the liquid-vapour interface. Let also j_c is a density of a molar vapour flux across the liquid–vapour interface, when a density of mass vapour flux across this interface is $j_m = j_c \cdot M$, where M is the molar mass of an evaporating substance (water). Let the normal velocity of the interface itself is u_Γ, see Fig. 2. Then the boundary condition for the normal velocity of liquid at the liquid–vapour interface reads:

$$\rho_l(\vec{u}_l \cdot \vec{n} - u_\Gamma) = j_m,$$

where ρ_l is the liquid density. Expressions for evaporation flux, j_m, and normal interfacial velocity, u_Γ, are specified below. A condition of the stress balance at the liquid–vapour interface is used to obtain boundary conditions for the pressure and tangential velocity:

$$(\mathbf{T} \cdot \vec{n})_l - (\mathbf{T} \cdot \vec{n})_g = -\gamma(\nabla \cdot \vec{n})_\Gamma \vec{n} + \gamma'_T(\nabla_\Gamma T), \quad (2)$$

where $\mathbf{T}$ is the full stress tensor; γ is the interfacial tension of the liquid–vapour interface, γ'_T is the derivative of the interfacial tension with the temperature; $\nabla_\Gamma T$ is the surface gradient of temperature, $(\nabla \cdot \vec{n})_\Gamma$ is the divergence of the normal vector at the liquid–vapour interface, which is equal to the curvature of the interface. Calculating a dot-product of (2) and the tangent vector $\vec{\tau}$ (see Fig. 2) and neglecting the viscous stress in the gas phase we deduce the boundary condition for thermal Marangoni convection, which determines the tangent component of the velocity vector. Calculating a dot-product of (2) and the normal vector $\vec{n}$ (see Fig. 2) and neglecting normal viscous stress in gas phase we obtain boundary condition for pressure in liquid at the liquid–vapour interface.

Since there is no penetration at the solid surface, the normal flux of vapour at the vapour–solid interface is zero:

$$\nabla c \cdot \vec{n} = 0,$$

where c is the molar concentration of the vapour in the air; and $\vec{n}$ is the unit vector, perpendicular to the solid–vapour interface.

Although we neglect convective terms in diffusion equations for the vapour phase, we leave vapour velocity at the derivation of the boundary conditions at the liquid–vapour interface in order to derive boundary conditions even for the case when the convection cannot be neglected. It is interesting that in the end the vapour velocity is cancelled out from the resulting boundary conditions for diffusion equation.

We consider the air phase as the mixture of vapour and dry air. Note that due to mass conservation law mass flux of vapour perpendicular to the liquid–vapour interface in the vapour phase should be equal to the mass flux of liquid perpendicular to the interface in the liquid phase:

$$\begin{aligned} j_m &= \rho_l(\vec{u}_l \cdot \vec{n} - u_\Gamma) \\ &= \rho_v(\vec{u}_g \cdot \vec{n} - u_\Gamma) - D_{vapour\ in\ air} \nabla \rho_v \cdot \vec{n}, \end{aligned} \quad (3)$$

Density of the mass flux of the dry air, $j_{m,\ air}$, in the air at the liquid–vapour interface:

$$j_{m,\ air} = \rho_{air}(\vec{u}_g \cdot \vec{n} - u_\Gamma) - D_{air\ in\ vapour} \nabla \rho_{air} \cdot \vec{n} = 0, \quad (4)$$

where ρ_v and ρ_{air} are densities of vapour and dry air, respectively; D is the diffusion coefficient; $\vec{u}_l$ and $\vec{u}_g$ are velocity vectors of liquid and gas, respectively; u_Γ and $\vec{n}$ are shown in Fig. 2. Note that flux of dry air across the interface is assumed to be zero. As the air under consideration includes more than one species of molecules, then the mass flux for each species in the air phase consists of two components: convective part, $\rho \vec{u}_g$, and diffusive one, $D\nabla\rho$. Flux in the pure liquid includes only a convective term, $\rho_l \vec{u}_l$. Fluxes are considered relatively to the liquid–vapour interface in the direction, normal to the interface. The latter results in an additional term: $-\rho u_\Gamma$. Let us adopt the following assumptions.

Assumption 1: $\rho_g = \rho_{air} + \rho_v = const$. This assumption results in

$$\nabla \rho_{air} = -\nabla \rho_v. \quad (5)$$

Assumption 2:

$$D_{air\ in\ vapour} = D_{vapour\ in\ air} = D. \quad (6)$$

After substitution of (5) and (6) into (3) and (4) and simple algebraic manipulations we arrive to an expression for the density of the mass flux across the liquid–vapour interface, j_m, as a function of molar vapour concentration, c, in the air:

$$j_m = \frac{-D\nabla\rho_v \cdot \vec{n}}{1 - \rho_v/\rho_g} = \frac{-D\nabla c \cdot \vec{n}}{1/M - c/\rho_g}, \quad (7)$$

where the following relation has been used: $\rho_v = cM$, M is the molar mass of an evaporating substance (water).

Equation 7 connects the evaporation flux, j_m, with both the gradient of vapour concentration in the normal direction and the concentration itself. On the other hand, the rate of mass transfer across the liquid–vapour interface is given by the Hertz-Knudsen-Langmuir equation [22, 23]:

$$j_m = \beta\sqrt{\frac{MRT}{2\pi}}(c_{sat}(T) - c), \qquad (8)$$

where β is the mass accommodation coefficient (probability that uptake of vapour molecules is occur upon collision of those molecules with the liquid surface); R is the universal gas constant; T and c are the local temperature in K° and molar vapour concentration at the liquid–vapour interface, respectively; c_{sat} is the molar concentration of saturated vapour. Molar concentration of saturated vapour is taken as a function of a local temperature and local curvature of the liquid–vapour interface according to Clausius-Clapeyron [24] and Kelvin [26] equations.

Combining (7) and (8) we obtain a boundary condition for diffusion equation (1) at the liquid–vapour interface:

$$\frac{-D\nabla c \cdot \vec{n}}{1/M - c/\rho_g} = \beta\sqrt{\frac{MRT}{2\pi}}(c_{sat}(T) - c).$$

At the axis of symmetry (r=0) the following boundary conditions are satisfied:

$$\left.\frac{\partial c}{\partial r}\right|_{r=0} = 0, \left.\frac{\partial T}{\partial r}\right|_{r=0} = 0, \left.\frac{\partial u_z}{\partial r}\right|_{r=0} = 0, \left.u_r\right|_{r=0} = 0, \left.\frac{\partial p}{\partial r}\right|_{r=0} = 0,$$

where u_r and u_z are radial and vertical components of the velocity vector, and p is the hydrodynamic pressure.

At the outer boundary of the computational domain values of temperature, T_∞, and vapour concentration, c_∞, are imposed.

In our computer simulations we assume that the droplet under consideration retains spherical-cap shape in the course of evaporation, and contact line is pinned (L=const). Then knowing the total mass evaporation flux, $J_m = \int_\Gamma j_m \, dA$ (dA is the element of area of the interface Γ), we can calculate the normal velocity of the liquid–vapour interface, u_Γ, at any point of the interface:

$$u_\Gamma = \frac{-J_m}{\pi\,\rho_l\,L^2} \cdot \frac{z}{(z + \Delta z) \cdot n_z + r \cdot n_r} \cdot \frac{1 + \cos\theta}{1 - \cos\theta},$$

where $\Delta z = L \cdot \cos\theta/\sin\theta$; θ is the contact angle; n_r and n_z are radial and vertical components of the vector $\vec{n}$, respectively, shown in Fig. 2; the origin of cylindrical coordinates (r, z) is supposed to be at the point of intersection of axis of droplet symmetry and the liquid–solid interface (Fig. 1).

Results and Discussion

Computer simulations are performed using commercial software COMSOL Multiphysics.

Computer simulations show that due to small size of the system under consideration diffusive fluxes from the liquid–vapour interface into the surrounding air dominate over convective ones. As the droplet size decreases, the domination of diffusion becomes stronger. Thus for water droplets of the size less than one micron the diffusive mass transfer (vapour in gas) is stronger than convective one by more than one order of magnitude. This result proves the validity of neglecting convective fluxes in gas phase and using only diffusion. Note for relatively big droplets (1 mm) the Peclet number is small and the convection can be neglected also [24].

The described above model has been validated against available experimental data [18] and results of computer simulations of diffusion limited evaporation [24], which were performed earlier for millimetre sized droplets. There is no disagreement with previous results [24] detected. Computer simulations show that the variation of mass accommodation coefficient, β, almost does not affect the value of evaporation flux for millimetre sized droplets, which confirms the insignificance of kinetic effects for millimetre sized droplets.

For a water droplet of one micron size the influence of mass accommodation coefficient, β, on the evaporation flux is already distinctly pronounced, which shows that kinetic effects come into play for droplets of submicron size.

For the rest of results presented here the value of β is taken as 0.5, which is the average experimentally measured value of β for water according to [22].

Comparison of the total evaporation flux, J_c, calculated with the present computer model and that calculated with the purely diffusive model of evaporation [24] shows that it is difficult to see any difference between these two values at the scale of droplet sizes ~1 mm.

It is well known that in the diffusive model of evaporation the total evaporation flux, J_c, is linearly proportional to the droplet size, L. However, according to the kinetic model of evaporation the evaporation process is limited by the rate of molecules transition from liquid phase to the gaseous one. That means that total evaporation flux, J_c, in kinetic model is proportional to the area of the liquid–vapour interface, or $J_c \sim L^2$. As we have already seen, the kinetic effects come into play when the size of the droplet is in the micrometers range. Therefore, in order to see the quadratic dependence ($J_c \sim L^2$)

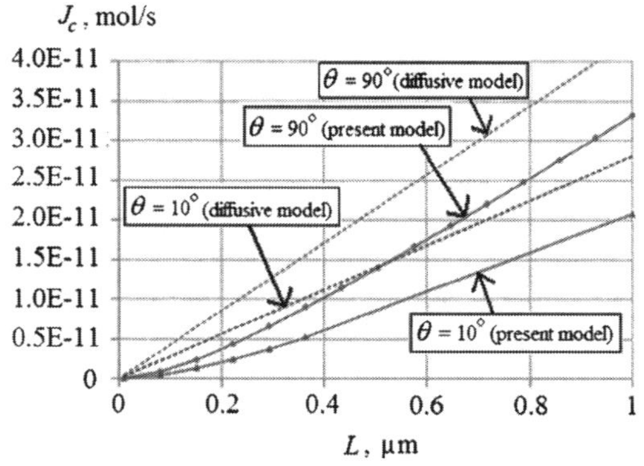

Fig. 3 Dependence of total molar evaporation flux, J_c, on the droplet size, L, for purely diffusive model of evaporation (*dashed lines*) [24], and for present computer model of evaporation (*solid lines*); β=0.5; relative air humidity is 70%

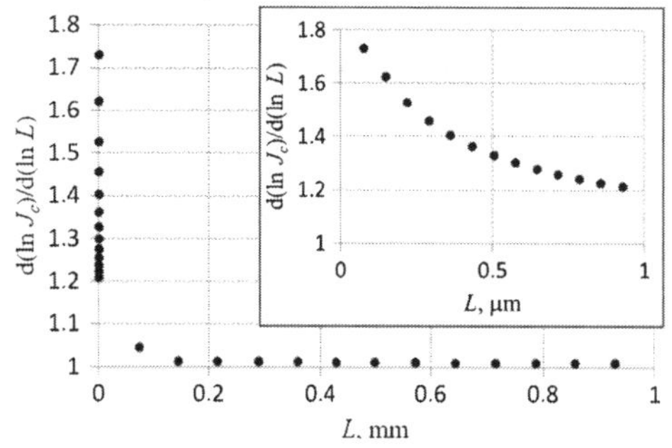

Fig. 4 Dependence of quantity $d(\ln J_c)/d(\ln L)$ on the droplet size, L, for the present model of evaporation; β=0.5; θ=90°; relative air humidity is 70%. The insertion shows the same dependence, but for $L <$ 1 µm

we have to select L less than 1 µm (see Fig. 3). These plots demonstrate that results of the present computer simulations are different from linear dependences which are typical for diffusive model of evaporation.

In order to check that obtained dependences represent smooth transition from quadratic dependence (kinetic model) to linear one (diffusive model) we plotted the following quantity: $d(\ln J_c)/d(\ln L)$, which represents the power n of the dependence $J_c \sim L^n$ (see Fig. 4). Figure 4 shows that the power n increases from 1.2 and approaches 2 as L decreases from 1 µm. The latter shows that J_c is tending to be proportional to L^2 as the size of the droplet decreases below 1 µm, which means that evaporation flux becomes proportional to the area of the liquid–vapour interface. On the other hand, while the size of a droplet increases from 1 µm to 1 mm, the power n reaches its stationary value of 1 (see Fig. 4), which indicates that J_c becomes proportional to the droplet perimeter, as it should be in a purely diffusive model of evaporation.

There is a low limit of L, where the above consideration can be applied. This limit is determined by the range of surface forces action, which is usually around 0.1 µm [20]. Below this limit the whole droplet is in the range of surface forces action and there is no spherical part any more even on the droplet's top (microdroplets according to [20]). Evaporation in the latter case should be very much different from the presented above.

Figure 5 shows the influence of kinetic effects on dependence of normalized evaporation rate, $J_c/J_c(\theta=90°)$, on contact angle, θ. In fact, the dependence for $L=10^{-3}$ m, presented in Fig. 5, coincides with the well-known dependence obtained by Picknett and Bexon [27] for isothermal diffusion-limited evaporation of sessile or pendant drops in still air. Dependence for $L=10^{-6}$ m is already slightly different from that for diffusion-limited evaporation. For yet smaller droplets ($L=10^{-8}$ m) this difference becomes

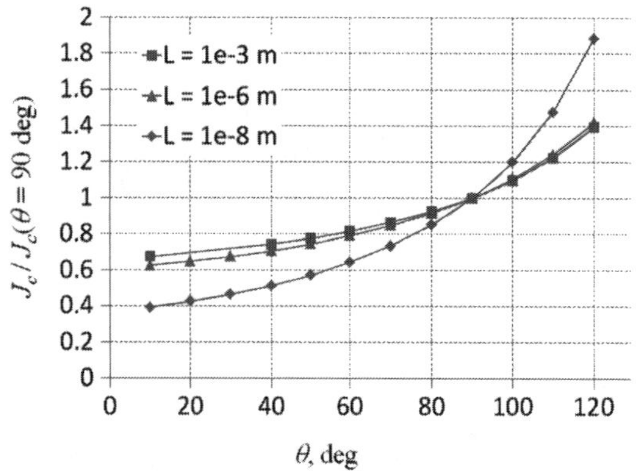

Fig. 5 Dependence of normalized evaporation rate on contact angle, θ. β=0.5; relative air humidity is 70%

distinctively noticeable. Thus, according to our computer simulations, the normalized evaporation rate, $J_c/J_c(\theta=90°)$, increases faster with increasing contact angle for smaller droplets as compared with bigger droplets.

Conclusion

Computer simulations of sessile small droplets of water are-performed. Present model combines diffusive and kinetic models of evaporation. Results of modelling show a smooth transition between different regimes of evaporation corresponding to diffusive and kinetic models of evaporation.

The dependence of droplet's evaporation rate on droplet size and contact angle is studied. It is shown that for large droplets (millimetre-size) the present model of evaporation is a pure diffusive model of evaporation. However, when the droplet size decreases down to around 1 µm or smaller, the evaporation is affected by kinetic processes, and kinetic theory at the liquid–vapour interface (Hertz-Knudsen-Langmuir equation) must be applied together with the diffusion equation of vapour in the air.

The vapour concentration at the liquid–vapour interface is in between its saturated value and its ambient value according to the presented model. The latter drives both the vapour diffusion and the process of molecules transfer from the liquid phase to the vapour phase. As a result the evaporation rate according to our model is always lower than the evaporation rate given independently by either pure diffusive or pure kinetic models of evaporation.

The presented model can be applied for evaporation of any other pure simple liquids, not water only.

Acknowledgements This research was supported by the European Union under Grant MULTIFLOW, FP7-ITN-2008-214919. V. M. Starov's research was also supported by the Engineering and Physical Sciences Research Council, UK (Grant EP/D077869/1). The work of R. G. Rubio was supported in part by the Spanish Ministerio de Ciencia e Innovación through grant FIS2009-14008-C02-01, and by ESA through project MAP-AO-00-052. Both V. M. Starov and R. G. Rubio recognise a support from European Space Agency (PASTA project).

References

1. Sodtke C, Stephan P (2007) Spray cooling on micro structured surfaces. Int J Heat Mass Tran 50:4089–4097
2. Cheng W-L, Han F-Y, Liu Q-N, Zhao R, Fan H-L (2011) Experimental and theoretical investigation of surface temperature non-uniformity of spray cooling. Energy 36:249–257
3. Du P, Li L, Zhao W, Leng X, Hu X (2011) Study on the printing performance of coated paper inkjet ink. Adv Mater Res 174:358–361
4. Campbell PG, Weiss LE (2007) Tissue engineering with the aid of inkjet printers. Expert Opin Biol Ther 7(8):1123–1127
5. Fuller SB, Wilhelm EJ, Jacobson JM (2002) Ink-jet printed nanoparticle microelectromechanical systems. J Microelectromech Syst 11:54–60
6. Haschke T, Wiechert W, Graf K, Bonaccurso E, Li G, Suttmeier FT (2007) Evaporation of solvent microdrops on polymer substrates: from well controlled experiments to mathematical models and back. Nanoscale Microscale Therm Eng 11:31–41
7. Pericet-Camara R, Bonaccurso E, Graf K (2008) Microstructuring of polystyrene surfaces with nonsolvent sessile droplets. Chemphyschem 9:1738–1746
8. Karlsson S, Rasmuson A, Björn IN, Schantz S (2011) Characterization and mathematical modelling of single fluidised particle coating. Powder Technol 207:245–256
9. Kim JH, Shi W-X, Larson RG (2007) Methods of stretching DNA molecules using flow fields. Langmuir 23:755–764
10. Deegan RD, Bakajin O, Dupont TF, Huber G, Nagel SR, Witten TA (2000) Contact line deposits in an evaporating drop. Phys Rev E 62:756–765
11. Guena G, Poulard C, Voue M, Coninck JD, Cazabat AM (2006) Evaporation of sessile liquid droplets. Colloid Surf A 291:191–196
12. Girard F, Antoni M, Sefiane K (2008) On the effect of Marangoni flow on evaporation rates of heated water drops. Langmuir 24:9207–9210
13. Hu H, Larson RG (2006) Marangoni effect reverses coffee-ring depositions. J Phys Chem B 110:7090–7094
14. Sefiane K, Wilson SK, David S, Dunn GJ, Duffy BR (2009) On the effect of the atmosphere on the evaporation of sessile droplets of water. Phys Fluids 21:062101
15. Ristenpart WD, Kim PG, Domingues C, Wan J, Stone HA (2007) Influence of substrate conductivity on circulation reversal in evaporating drops. Phys Rev Lett 99:234502
16. Bhardwaj R, Fang X, Attinger D (2009) Pattern formation during the evaporation of a colloidal nanoliter drop: a numerical and experimental study. New J Phys 11:075020
17. David S, Sefiane K, Tadrist L (2007) Experimental investigation of the effect of thermal properties of the substrate in the wetting and evaporation of sessile drops. Colloid Surf A 298:108–114
18. Semenov S, Starov VM, Rubio RG, Agogo H, Velarde MG (2011) Evaporation of sessile water droplets: universal behaviour in presence of contact angle hysteresis. Colloid Surf A. doi:10.1016/j.colsurfa.2011.07.013, in press
19. Moosman S, Homsy GM (1980) Evaporating menisci of wetting fluids. J Colloid Interface Sci 73:212–223
20. Starov V, Velarde M, Radke C (2007) Dynamics of wetting and spreading. In: Surfactant sciences series, vol 138. Taylor & Francis Boca Raton
21. Ajaev VS, Gambaryan-Roisman T, Stephan P (2010) Static and dynamic contact angles of evaporating liquids on heated surfaces. J Colloid Interface Sci 342:550–558
22. Kryukov AP, Levashov VYu, Sazhin SS (2004) Evaporation of diesel fuel droplets: kinetic versus hydrodynamic models. Int J Heat Mass Tran 47:2541–2549
23. Sazhin SS, Shishkova IN, Kryukov AP, Levashov VYu, Heikal MR (2007) Evaporation of droplets into a background gas: kinetic modelling. Int J Heat Mass Tran 50:2675–2691
24. Semenov S, Starov VM, Rubio RG, Velarde MG (2010) Instantaneous distribution of fluxes in the course of evaporation of sessile liquid droplets: computer simulations. Colloid Surf A 372:127–134
25. Bligh PH, Haywood R (1986) Latent heat – its meaning and measurement. Eur J Phys 7:245–251
26. Galvin KP (2005) A conceptually simple derivation of the Kelvin equation. Chem Eng Sci 60:4659–4660
27. Picknett RG, Bexon R (1977) The evaporation of sessile or pendant drops in still air. J Colloid Interface Sci 61:336–350

Progr Colloid Polym Sci (2012) 139
DOI: 10.1007/978-3-642-28974-3_15

Viscosity of Rigid and Breakable Aggregate Suspensions Stokesian Dynamics for Rigid Aggregates

R. Seto[1], R. Botet[2], and H. Briesen[1]

Abstract Suspensions of rigid aggregates have been investigated by Stokesian dynamics. In our recent work (Seto R, Botet R, Briesen H. Phys Rev E 84:041405, 2011), the motions of freely suspended aggregates in shear flows were determined by considering the force and torque balance, and forces and moments acting on the contact points within aggregates were also evaluated. Here, by comparing the obtained results with a bond strength between particles, the sustainable sizes of the aggregates under shear flows have been estimated, which leads to a viscosity-shear rate relation. Our method allows us to see not only the power-law shear thinning for fractal aggregates but also some deviations due to the finite-size effects.

Introduction

Particles suspended in a fluid affect the viscosity. Fluid flow is disturbed by such particles, which causes an increase of the viscosity. If particles are individually dispersed, the viscosity of the dilute limit of the suspension depends only on the volume fraction of the particles. If the particles are aggregated, the viscosity depends on the sizes and structures of the aggregates as well. We consider the situation where the total number of primary particles in the system remains constant. In this case, the flow disturbance per aggregate increases due to its increasing size, while the number density of aggregates decreases. Therefore, the apparent viscosity is determined by the efficiency of the structure to disturb flow. Furthermore, the aggregates can be broken or restructured by an applied flow [1–11], which eventually may cause non-Newtonian behaviour.

When particles interact with each other by only hydrodynamic forces, the shear-rate dependence of suspension's viscosity is explained by the contribution of Brownian motion and the formation of microstructures [12–14]. In this case, the ratio between applied flow and thermal agitations, i.e. Péclet number, is the major parameter to determine the behaviour. However, in this paper, we consider a different situation in which particles are aggregated due to bonding forces, and hydrodynamic forces may break them up. Here, the contribution of Brownian motion is neglected because it is less important for larger aggregates at lower shear rates. In this case, the competition between hydrodynamic force and bond strength is expected to determine the sustainable size of aggregates and as a result the viscosity. In general, the estimation of the sustainable size of aggregates under flows is not simple. In our previous work, the restructuring of colloidal aggregates has been studied by DEM simulation with the free-draining approximation [15–17]. There it has been shown that aggregates feature a rigid rotation regime under a lower shear-rate, and a restructuring regime above a certain shear rate. If new bonds are generated between newly contacting particles, the aggregates can become more robust by forming multi-linking structures. However, in this work we focus on hydrodynamic interactions, so a highly idealized condition is assumed to determine the sustainable size of aggregates. Shear thinning due to the breakup of fractal aggregates has already been shown by Wessel and Ball [18]. They applied Einstein's equation to aggregates suspensions using the radii of gyration R_g of fractal aggregates in place of radii of spheres. By estimating the sustainable maximum sizes of brittle aggregates under shear flows, they found a power-law dependence in the increment of the relative viscosity for the shear rate, i.e. $\Delta\eta_r \sim \dot{\gamma}^{-(3-D_f)/3}$.

In this work, we have studied the dependence of the viscosity on structure and size of aggregates by employing Stokesian dynamics (SD) [19–21] for more

R. Seto (✉)
[1]Process Systems Engineering, Technische Universität München, Weihenstephaner Steig 23, Freising D-85350, Germany
e-mail: setoryohei@me.com
[2]Laboratoire de Physique des Solides, Université Paris-Sud, UMR8502, Bât 510, Orsay 91405, France

detailed hydrodynamic interactions. By using the force-torque-stresslet (FTS) version of SD, the forces, torques, and stresslets acting on particles are approximately related to the velocities and angular velocities in a shear flow. In our recent work, we showed a method to determine the motions of force- and torque-free rigid aggregates in shear flows without simulating their time-evolution [22]. Though this method cannot deal with deformation or restructuring of aggregates, it allows one to evaluate the hydrodynamic response of specific sizes and structures with little computational effort. In this work, we have extend the method to evaluate the viscosity of aggregate suspensions by using the relation between stress and viscosity given by Batchelor [23]. As a demonstrative example, we have used a simple breakup criteria where the forces and moments acting on the contact points within aggregates are compared to a bond strength between particles. Using this simple criterion is a strong idealization especially for a moment load at the contact points. Instead of instant breakup, mutual rolling may be more appropriate [24–26]. However, the benefit of the presented approach is that it avoids time-evolution simulations which should be necessary for investigating mutual rolling.

The characterisation of contact forces is one of the important issues in colloidal science, so many observations have been reported over the recent years [27–31]. We have evaluated the sustainable size of fractal-like aggregates by using an experimental result from the literature and determined the shear-rate dependence of their viscosities.

Viscosity of Particle Suspension

The viscosity η_0 of a Newtonian fluid is seen in the relation between shear stress τ and shear rate $\dot{\gamma}$ for simple shear-flow conditions, i.e. $\tau = \eta_0\dot{\gamma}$. If particles are dispersed in the fluid, the flow disturbances caused by them result in an increase of the viscosity. When interparticle interaction and Brownian motion can be neglected, the influence of the particles appears as an additional term nS [23];

$$\tau = \eta_0\dot{\gamma} + nS \tag{1}$$

where S is strength of the stresslet tensor S acting on a particle, and n is number density of the particles. By using the relative viscosity η_r, the apparent viscosity is given as $\eta = \eta_0\eta_r$. So, the increment of η_r is written as

$$\Delta\eta_r \equiv \eta_r - 1 = \frac{nS}{\eta_0\dot{\gamma}}. \tag{2}$$

For the dilute limit, the stresslet S can be evaluated by the analytical solution of the Stokes equations for a single-sphere system. When a sphere moves with a velocity $\boldsymbol{U}$ and angular velocity $\boldsymbol{\Omega}$ in a linearly changing flow $\boldsymbol{U}^{\infty}(\boldsymbol{r}) = \boldsymbol{U}^{\infty} + \boldsymbol{\Omega}^{\infty} \times \boldsymbol{r} + \mathsf{E}^{\infty}\boldsymbol{r}$, the hydrodynamic interactions, i.e. the drag force $\boldsymbol{F}$, torque $\boldsymbol{T}$ and stresslet S, are given by

$$\begin{aligned} \boldsymbol{F} &= -6\pi\eta_0 a(\boldsymbol{U} - \boldsymbol{U}^{\infty}(\boldsymbol{r})), \\ \boldsymbol{T} &= -8\pi\eta_0 a^3(\boldsymbol{\Omega} - \boldsymbol{\Omega}^{\infty}), \\ \mathsf{S} &= \frac{20}{3}\pi\eta_0 a^3\mathsf{E}^{\infty}, \end{aligned} \tag{3}$$

with first relation being known as Stokes' law.

Since the inertia of colloidal particles is negligibly small and the kinetic energy of particles is easily dispersed by the viscous solvent, the force and torque are minimized by the motion, i.e. the translation and rotation just follow the imposed flow;

$$(\boldsymbol{U},\ \Omega) \to (\boldsymbol{U}^{\infty}(\boldsymbol{r}),\ \Omega^{\infty}) \Rightarrow (\boldsymbol{F},\boldsymbol{T}) \to (\boldsymbol{0},\boldsymbol{0}). \tag{4}$$

The stresslet S is not cancelled by any motion and remains finite. In a simple shear flow $\boldsymbol{U}^{\infty}(\boldsymbol{r}) = \dot{\gamma}z\boldsymbol{e}_x$, the non-zero elements of stresslet are given by $S_{xz} = (10/3)\pi\eta_0 a^3\dot{\gamma}$. By substituting this result into (2), the standard Einstein's equation is obtained:

$$\Delta\eta_r = \frac{10}{3}\pi n a^3 = \frac{5}{2}\phi, \tag{5}$$

where the volume fraction $\phi \equiv n \times 4/3\pi a^3$ is used for the last expression.

Viscosity of Aggregate Suspension

If particles are aggregated, Einstein's equation (5) is not applicable. Here, we consider the dilute limit of strongly aggregated suspensions, i.e. primary particles are rigidly connected within an aggregate, and the distance between the nearest aggregate is so large that the hydrodynamic interaction can be evaluated for isolated aggregate.

The increment of the relative viscosity for aggregate suspensions can be written as follows;

$$\Delta\eta_r = \frac{n'\langle S_{ag}\rangle}{\eta_0\dot{\gamma}}, \tag{6}$$

where $\langle S_{ag}\rangle$ is the averaged value of the stresslets acting on single aggregates, and n' is the number density of the aggregates. When the structures of the aggregates are assumed to be rigid, the total force $\boldsymbol{F}_{\text{ag}}$, torque $\boldsymbol{T}_{\text{ag}}$ and stresslet $\boldsymbol{S}_{\text{ag}}$[1] are given by a 11×11 resistance matrix R_{ag} [22, 32];

$$\begin{pmatrix} \boldsymbol{F}_{\text{ag}} \\ \boldsymbol{T}_{\text{ag}} \\ \boldsymbol{S}_{\text{ag}} \end{pmatrix} = -\mathsf{R}_{\text{ag}} \begin{pmatrix} \boldsymbol{U}_{\text{ag}} - \boldsymbol{U}^{\infty}(\boldsymbol{r}_0) \\ \boldsymbol{\Omega}_{\text{ag}} - \boldsymbol{\Omega}^{\infty} \\ -\boldsymbol{E}^{\infty}_{\text{ag}} \end{pmatrix}. \tag{7}$$

The matrix R_{ag} can be determined from the grand resistance matrix of SD with the constraint of the rigid structure. Since the relative velocities between any two particles within a rigid aggregate are zero, the lubrication correction of SD are less important. But, it rather can cause errors due to the two-body solution applied at the inside of the aggregate, where the external flow can be weakened by the particles [33]. This is why, the lubrication correction is omitted in this work.

We consider the small-inertia limit to estimate the motion of the aggregates. There, $(\boldsymbol{F}_{\text{ag}}, \boldsymbol{T}_{\text{ag}})=(\boldsymbol{0},\boldsymbol{0})$ should be satisfied as in the previous section. By solving (7) with the force- and torque-free conditions, the motion of an aggregate ($\boldsymbol{U}_{\text{ag}}$, $\boldsymbol{\Omega}_{\text{ag}}$) is determined [22]:

$$\begin{aligned} \boldsymbol{U}_{\text{ag}} &= \boldsymbol{U}^{\infty}(\boldsymbol{r}_0) - \left(\mathsf{R}^{-1}_{\text{ag}}\right)_{\text{US}} \boldsymbol{S}_{\text{ag}}, \\ \boldsymbol{\Omega}_{\text{ag}} &= \boldsymbol{\Omega}^{\infty} - \left(\mathsf{R}^{-1}_{\text{ag}}\right)_{\Omega\text{S}} \boldsymbol{S}_{\text{ag}}, \end{aligned} \tag{8}$$

where the subscripts US and ΩS indicate the submatrices of the inverse of the resistance matrix that relate $\boldsymbol{U}$ and $\boldsymbol{S}$, and $\boldsymbol{\Omega}$ and $\boldsymbol{S}$, respectively. The motion given by (8) is referred to as the *torque-balanced motion*. Simultaneously, the stresslet $\boldsymbol{S}_{\text{ag}}$ acting on the aggregate is also determined;

$$\boldsymbol{S}_{\text{ag}} = \left(\left(\mathsf{R}^{-1}_{\text{ag}}\right)_{\text{ES}}\right)^{-1} \boldsymbol{E}^{\infty}. \tag{9}$$

We studied different types of randomly structured aggregates. The clusters are generated by three types of numerical models, i.e. reaction-limited hierarchical cluster-cluster aggregation (CCA), diffusion-limited particle-cluster aggregation (DLA) and the Eden growth model (Eden) [34]. At the large limit, the fractal dimensions D_{f} of the generated aggregates expected to be 2.04, 2.5 and 3, respectively. Figure 1 shows the averages and standard deviations of the radius of gyration R_{g}.[2] In this paper, the averages and standard deviations have been taken at the log scale over 256 samples at each size $N=2^m$, $m=3, \ldots, 10$. As seen in the plots, CCA clusters follow a power law $R_{\text{g}} \propto N^{1/d_{\text{f}}}$, and the fitting value

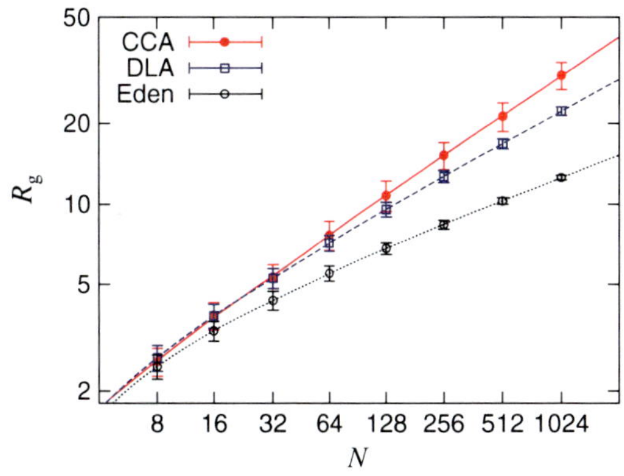

Fig. 1 N dependence of the radius of gyration R_{g} for CCA (*red disk* and *solid line*), DLA (*blue box* and *dashed line*), and Eden (*black circle* and *dotted line*) clusters. The averages and standard deviations are taken over 256 samples

of $d_{\text{f}} \approx 2$ is close to the literature value [34], i.e. $D_{\text{f}} \approx 2.04$. However, for small sizes DLA and Eden clusters show some deviations from power-law behaviour. The DLA and Eden models yield more open-structured configurations. For small clusters, this effect cannot be neglected, so that the estimated fractal dimensions appear to be smaller. In order to deal with fractal-like clusters with such a finite-size effect, a power-law relation cannot be assumed *a priori*.

Here, an example is shown as a demonstration for the torque-balanced motion. Figure 2 represents the drag forces acting on particles within a CCA cluster of $N=256$, which is the projections for the 3D positions and forces of the particles in the xz plane. In this work, we have used the numerical library developed by Ichiki [35] for obtaining it.

The N dependence of the stresslet for the three types of aggregates have been obtained as shown in Fig. 3, where the strength S_{ag} is obtained by $S^2_{\text{ag}} \equiv (1/2)\sum_{j,k} S^2_{jk}$. It is worth noting that the stresslet growth for the CCA clusters follows the expected relation; $S_{ag} \approx (10/3)\,\pi\eta_0\dot{\gamma}R^3_{\text{g}}$. In any cases, three-parameter fittings with (λ, k, l) have been applied for the average values in order to capture the deviations from the asymptotic behaviour for small N:

$$S_{\text{ag}} = \frac{10}{3}\pi\eta_0 a^3 \dot{\gamma} k N^{\lambda}\, e^{l/\log N}, \tag{10}$$

This functional form corresponds to the use of $\log S_{\text{ag}}=\lambda \log N + C + l/\log N$ instead of the linear form $\log S_{\text{ag}}=\lambda \log N + C$. The averages have been taken over 256 samples at each size, but not over the different orientations for each sample. The obtained interpolating functions are shown by the lines in Fig. 3, and the values of the fitting parameters obtained are given in Table 1.

We consider a situation where the number density of the primary particles n is conserved in a suspension. Then the

[1] The bold italic symbol for stresslet implies the five independent elements of the symmetric traceless tensor: $\boldsymbol{S} \equiv (S_{xx}, S_{xy}, S_{xz}, S_{yz}, S_{yy})$.

[2] The radius of gyration is defined by $R^2_{\text{g}} \equiv (1/N)\sum_i (r^{(i)} - r_0)^2$

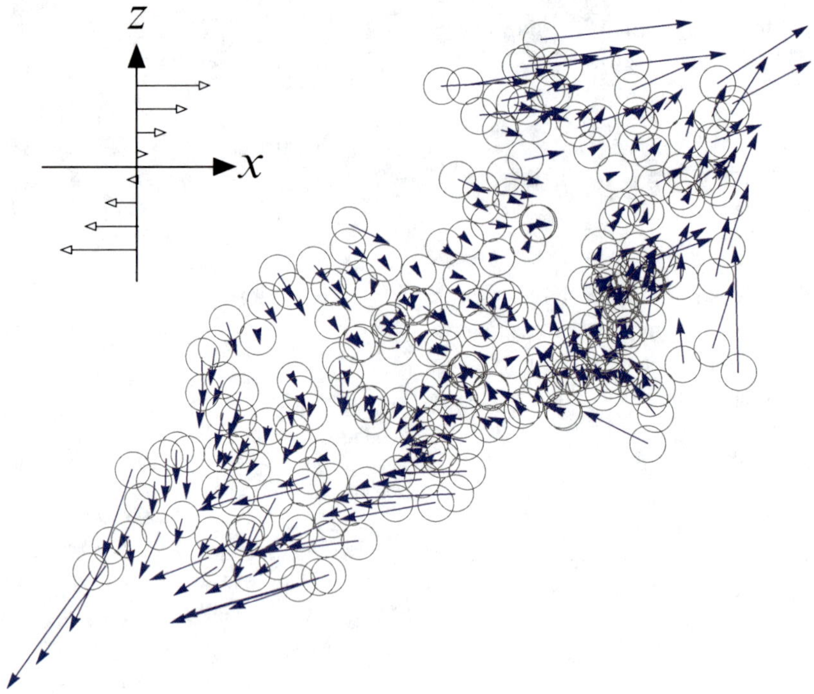

Fig. 2 The *arrows* show the drag forces acting on the primary particles within a CCA cluster (N=256) at the torque-balanced motion. The projected spheres on xz plane appear to overlap, but they do not in the original 3D structure indeed

number density of aggregates n' satisfies $n = \langle N \rangle n'$, where $\langle N \rangle$ is the average number size of aggregates. If the suspension can be considered as consisting of monodisperse aggregates i.e. $\langle N \rangle = N$, the N dependence of $\Delta\eta_{\mathrm{r}}$ is obtained as follows:

$$\Delta\eta_{\mathrm{r}} \approx \frac{n\langle S_{\mathrm{ag}}\rangle}{N\,\eta_0\dot{\gamma}} \approx \frac{5}{2}\phi k N^{\lambda-1} e^{l/\log N} \tag{11}$$

The viscosity does not depend on the shear rate as long as the size of aggregates is unchanged.

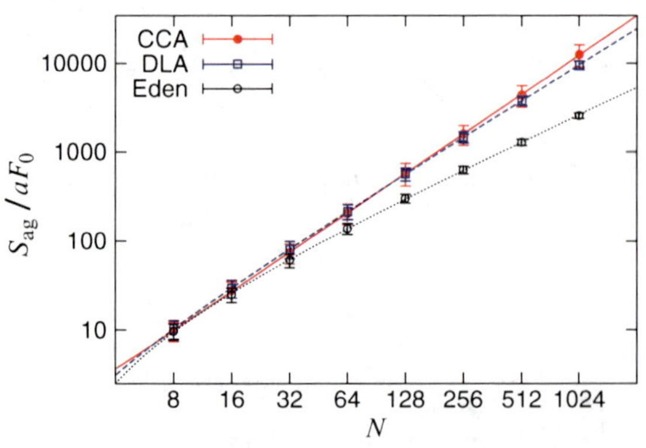

Fig. 3 Strengths of the stresslet tensor $S_{\mathrm{ag}} \equiv \left((1/2)\sum_{j,k} S_{jk}^2\right)^{1/2}$ for CCA (*red disk* and *solid line*), DLA (*blue box* and *dashed line*), and Eden (*black circle* and *dotted line*) clusters are shown, which is normalized by $aF_0 = 6\pi\eta_0\dot{\gamma}a^3$. The lines represent the fit of (10) to the values

Shear Thinning of Rigid and Breakable Aggregate Suspensions

In this section, we consider a possible scenario to demonstrate shear-thinning behaviour by introducing some assumptions. We have examined the disturbance of flows by aggregates of a certain size and structure and the viscosities of their suspensions as a result. Conversely, hydrodynamic forces can cause a change in aggregate size and structure. The determination of final size and structure under a shear flow requires a time-evolution simulation by coupling discrete element method and Stokesian dynamics, which is out of the scope of this paper. Here, we only consider the an idealized case where the size is determined by simple criteria and the structure remains unchanged.

If an aggregate does not have any no loop structure, the forces and moments acting on the contact points can be determined by considering the equilibrium conditions [22]. The number of the contact points within aggregates generated by the fractal models is always $N-1$, so no loop can be formed. Even if it had a few loops, a single linked junction may cause breakup of the aggregate [36]. In that case, the following argument is applicable.

The total force and torque are zero for the torque-balanced motion, so the equilibrium conditions for the forces and moments acting on each contact point $\alpha = 1, \ldots, N-1$ are

Table 1 Fitting parameters for the N dependence of the stresslet S_{ag} given by (10)

	λ	k	l
CCA	1.48	0.75	0.16
DLA	1.33	2.01	−1.14
Eden	0.97	8.49	−2.74

satisfied. Thus, the forces and moments can be determined. The obtained results are decomposed into the four mechanical modes; normal elongation force $F^{(\alpha)}_{\mathrm{elong}}$, sliding force $F^{(\alpha)}_{\mathrm{slid}}$, bending moment $M^{(\alpha)}_{\mathrm{bend}}$, and torsional moment $M^{(\alpha)}_{\mathrm{torsion}}$. Only the elongation forces and bending moments have been considered here, because experimental data are only available for these two modes [30, 31]. When the maximum force or moment reaches the bond strength, the most stressed connection is expected to be broken. Therefore, the maximum values are important to estimate the sustainable size of aggregate under a shear flow. The N dependence of the maximum bending moment is shown in Fig. 4. The averaged values of $\max(F_{\mathrm{elong}})$ and $\max(M_{\mathrm{bend}})$ for N are fitted by the following relations with three fitting parameters (ζ, B, C);

$$\left\langle \max_{\alpha} F^{(\alpha)}_{\mathrm{elong}}/F_0 \right\rangle = C_{\mathrm{elong}} N^{\zeta_{\mathrm{elong}}} e^{B_{\mathrm{elong}}/\log N}, \qquad (12)$$

$$\left\langle \max_{\alpha} M^{(\alpha)}_{\mathrm{bend}}/aF_0 \right\rangle = C_{\mathrm{bend}} N^{\zeta_{\mathrm{bend}}} e^{B_{\mathrm{bend}}/\log N}. \qquad (13)$$

We introduce the critical number-size $N_c(\dot{\gamma})$. Under a shear flow of shear rate $\dot{\gamma}$, the aggregates larger than $N_c(\dot{\gamma})$ are broken, but the smaller ones remain. The criteria of $N_c(\dot{\gamma})$ is given by comparing the maximum force and moment for contact points with a given bond strength. The bond strength is assumed to be characterized by a critical elongation force F_{Nc} and a critical bending moment M_{Bc}. So, $N_c(\dot{\gamma})$ is determined by

$$N_c(\dot{\gamma}) = \min\left(N_c^{\mathrm{elong}}, N_c^{\mathrm{bend}}\right) \qquad (14)$$

where N_c^{elong} and N_c^{bend} satisfy the following equality, respectively;

$$\left\langle \max_{\alpha} F^{(\alpha)}_{\mathrm{elong}}/F_0 \right\rangle \left(N_c^{\mathrm{elong}}\right) = F_{\mathrm{Nc}}/F_0, \qquad (15)$$

$$\left\langle \max_{\alpha} M^{(\alpha)}_{\mathrm{bend}}/aF_0 \right\rangle \left(N_c^{\mathrm{bend}}\right) = M_{\mathrm{Bc}}/aF_0. \qquad (16)$$

We need an additional assumption for the relationship between the critical number-size $N_c(\dot{\gamma})$ and the average size of aggregates $\langle N \rangle$ in a sheared suspension. The average size $\langle N \rangle$ is assumed to be proportional to the critical number-size $N_c(\dot{\gamma})$, so that $\langle N \rangle = hN_c(\dot{\gamma})$. The proportionality

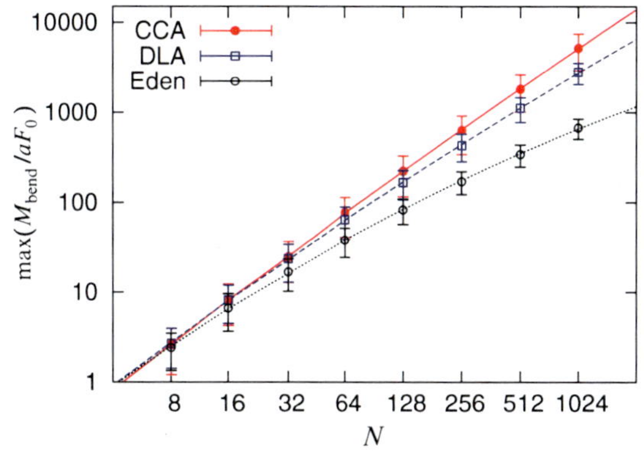

Fig. 4 The N dependence of the maximum bending moment for CCA (*red disk* and *solid line*), DLA (*blue box* and *dashed line*), and Eden (*black circle* and *dotted line*) clusters are shown, which are normalized by $aF_0 = 6\pi\eta_0 a^3\dot{\gamma}$. The interpolating lines are given by (13)

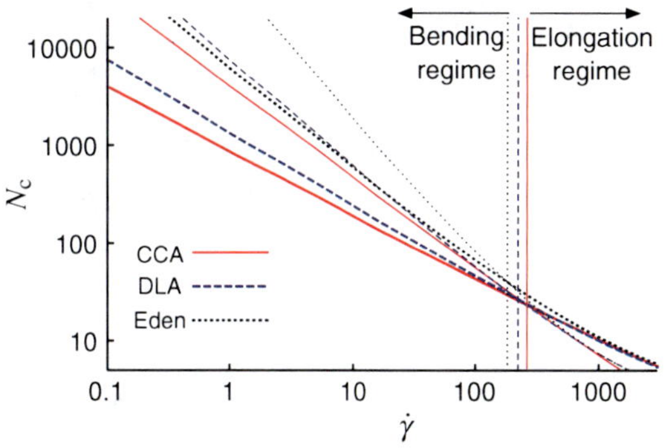

Fig. 5 The shear-rate dependence of the critical-number size $N_c(\dot{\gamma})$ are plotted for CCA (*red solid line*), DLA (*blue dashed line*), Eden (*black dotted line*). The *thin lines* are given by the force criterion (15) with $F_{\mathrm{Nc}} = 10$ pN, and the *thick lines* are given by the moment criterion (16) with $M_{\mathrm{Bc}} = 30$ pN μm. The cross points between two regimes are indicated by the *vertical lines*

constant h depends on the breakup and aggregation mechanism, i.e. restructuring and bonding condition, and so forth. In this work h is assumed to be unity. This considers the idealized case that fragmentation and restructuring processes eventually result in aggregates with just the critical size.

Finally, the shear-rate dependence of the relative viscosity η_r can be obtained. The bond strength is taken from experimental observations by using optical tweezers [30, 31], i.e. $F_{\mathrm{Nc}} = 10$ pN and $M_{\mathrm{Bc}} = 30$ pN·μm for $a = 735$ nm of PMMA particles. The solvent is water, so the viscosity and density are $\eta_0 = 10^{-3}$ Pa·s and $\rho_0 = 1{,}000$ kg/m^3, respectively. We have examined the range of shear rate from $\dot{\gamma} = 0.1\mathrm{s}^{-1}$ to $2000\mathrm{s}^{-1}$, which corresponds to the particle Reynolds number $\mathrm{Re} \equiv \rho_0 a^2\dot{\gamma}/\eta_0$ from 10^{-7} to 10^{-3}. As seen in Fig. 5, the critical-number size decreases monotonously with the shear rate. At the highest shear-rate, the critical-number size reach to the order of 10, which corresponds to

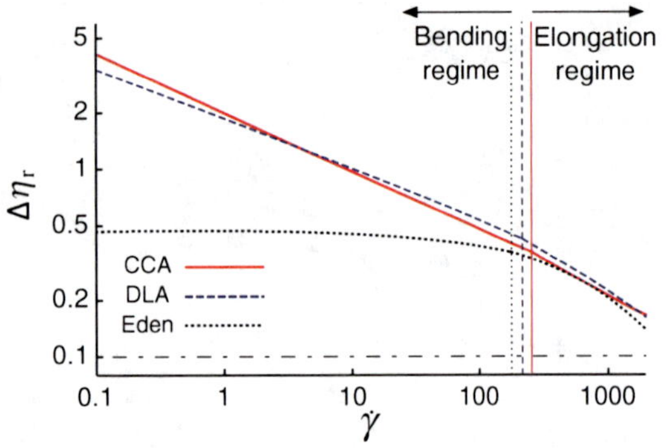

Fig. 6 The shear-rate dependence of $\Delta\eta_r$ are plotted for CCA (*red solid line*), DLA (*blue dashed line*) and Eden (*black dotted line*). The transitions between two regimes are shown by the *vertical lines*. The viscosity of the dispersed suspension with the same volume fraction is $\Delta\eta_r$=0.1, which is shown by the *dot-dashed line*

the minimum size used in this work. For larger aggregates, the breakups are caused by the bending-moment criteria (15), but for smaller aggregate they are caused by the elongation-force criteria (16). The volume fraction is chosen φ=0.04, which leads to $\Delta\eta_r$=0.1 for the dispersed system [see (5)]. By using (11), the expected value of $\Delta\eta_r$ are obtained as shown in Fig. 6. The open structured large aggregates cause a large increase of the viscosity. But, such aggregates are fragmented to smaller and smaller sizes by increasing the shear rate, which shows a shear-thinning behaviour. The shear thinning of CCA clusters follows a power law well and the exponent was about −0.3, which agrees with the result of Wessel and Ball [18], i.e. $-(3-D_f)/3\approx-0.3$ for $D_f\approx2.04$. For Eden clusters, the Newtonian behaviour predicted by $-(3-D_f)/3\approx0$ for $D_f\approx3$ is also seen at the lower shear-rate but it ends up with the shear thinning at the high shear rate which is brought by elongation forces. But it is worth noting that, even if the bending criteria remains at the higher shear rate, a similar shear thinning behaviour is seen, which is explained by the less compactness of smaller Eden clusters. The shear-rate dependence for N_c (Fig. 5) and $\Delta\eta_r$ (Fig. 6) were demonstrated by choosing a particular set of parameters here. More generic plots might be obtained by introducing a dimensionless quantity representing the ratio between shear rate and bond strength, such as $6\pi\eta_0 a^2\dot{\gamma}/F_{\mathrm{Nc}}$ or $6\pi\eta_0 a^3\dot{\gamma}/M_{\mathrm{Bc}}$. However, as we used a particular experimental parameter set from [30, 31], we limit ourselves to the particular case.

In the above evaluation, rigidity is assumed for the structure of aggregates, which maximizes the flow disturbance. However, for larger aggregates, elastic or/and plastic deformation can reduce the viscosity increase. Such deformation may also cause the consolidation of the aggregates, which requires more precise breakup criteria. Moreover, the idealized breakup criteria cannot capture the case that the aggregate are not split into two part by breakup of one bond due to the steric hindrance, which is expected for more compact clusters.

Acknowledgements The authors would like to thank Dr. K. Ichiki for providing the simulator of Stokesian dynamics "Ryuon" and many instructive advice, Dr. A. Braun for useful discussions, Dr. A. M. Reilly for proofreading of this manuscript, and Prof. R. Buscall for helpful suggestions. We also acknowledge the financial support of the German Science Foundation (DFG priority program SPP 1273, BR-2035/3-1,2).

References

1. Sonntag RC, Russel WB (1986) J Colloid Interface Sci 113(2):399
2. Serra T, Casamitjana X (1998) J Colloid Interface Sci 206:505
3. Selomulya C, Amal R, Bushell G, Waite TD (2001) J Colloid Interface Sci 236:67
4. Selomulya C, Bushell G, Amal R, Waite TD (2004) Int J Miner Process 73(2–4):295
5. Tolpekin VA, Duits MHG, van den Ende D, Mellema J (2004) Langmuir 20:2614
6. Soos M, Sefcik J, Morbidelli M (2006) Chem Eng Sci 61(8):2349
7. Soos M, Moussa AS, Ehrl L, Sefcik J, Wu H, Morbidelli M (2008) J Colloid Interface Sci 319:577
8. Ehrl L, Soos M, Morbidelli M (2008) Langmuir 24:3070
9. Zaccone A, Soos M, Lattuada M, Wu H, Babler MU, Morbidelli M (2009) Phys Rev E 79:061401
10. Frappier G, Lartiges BS, Skali-Lami S (2010) Langmuir 26 (13):10475
11. Harshe YM, Lattuada M, Soos M (2011) Langmuir 27:5739
12. Bossis G, Brady JF (1989) J Chem Phys 91:1866
13. Phung TN, Brady JF (1996) J Fluid Mech 313:181
14. Wagner NJ, Brady JF (2009) Phys Today 62:27
15. Becker V, Briesen H (2008) Phys Rev E 78:061404
16. Becker V, Schlauch E, Behr M, Briesen H (2009) J Colloid Interface Sci 339:362
17. Becker V, Briesen H (2010) J Colloid Interface Sci 346:32
18. Wessel R, Ball RC (1992) Phys Rev A 46(6):3008
19. Brady JF, Bossis G (1988) Ann Rev Fluid Mech 20:111
20. Durlofsky L, Brady JF, Bossis G (1987) J Fluid Mech 180:21
21. Ichiki K (2002) J Fluid Mech 452:231
22. Seto R, Botet R, Briesen H (2011) Phys Rev E 84:041405
23. Batchelor GK (1970) J Fluid Mech 41(3):545
24. Johnson KL (1985) Contact mechanics. Cambridge University Press, Cambridge
25. Dominik C, Tielens AGGM (1997) Astrophys J 480:647
26. Gastaldi A, Vanni M (2011) J Colloid Interface Sci 357:18
27. Ducker WA, Senden TJ, Pashley RM (1991) Nature 353:239
28. Heim LO, Blum J, Preuss M, Butt HJ (1999) Phys Rev Lett 83 (16):3328
29. Ecke S, Raiteri R, Bonaccurso E, Reiner C, Deiseroth HJ, Butt HJ (2001) Rev Sci Instrum 72(11):4164–4170
30. Pantina JP, Furst EM (2004) Langmuir 20(10):3940
31. Pantina JP, Furst EM (2005) Phys Rev Lett 94:138301
32. Harshe YM, Ehrl L, Lattuada M (2010) J Colloid Interface Sci 352:87
33. Bossis G, Meunier A, Brady JF (1991) J Chem Phys 94(7):5064
34. Jullien R, Botet R (1987) Aggregation and fractal aggregates. World Scientific, Singapore
35. Ichiki K (2011) Ryuon – simulation library for Stokesian dynamics. URL http://ryuon.sourceforge.net
36. Doi M, Chen D (1989) J Chem Phys 90(10):5271

Progr Colloid Polym Sci (2012) 139
DOI: 10.1007/978-3-642-28974-3_16

Neutron Reflection at the Calcite-Liquid Interface

Isabella N. Stocker[1], Kathryn L. Miller[1], Seung Y. Lee[1], Rebecca J.L. Welbourn[1], Alice R. Mannion[1], Ian R. Collins[2], Kevin J. Webb[2], Andrew Wildes[3], Christian J. Kinane[4], and Stuart M. Clarke[1]

Abstract The calcite-liquid interface is of great importance in many industrial and academic situations. In this paper, this interface is investigated using the powerful method of neutron reflection. Experimental approaches to overcome the challenges of surface roughness, cleanliness and mineral dissolution have been developed. Based on this protocol, the interfaces of calcite with the liquids water, toluene and heptane have been characterised successfully. Hence this study allows the technique of neutron reflection to be expanded to the investigation of mineral surfaces.

Introduction

Calcium carbonate ($CaCO_3$) is of great industrial importance owing to its abundance and non-toxicity. Applications include paper whitening and lubrication. Prevention of scale formation in domestic and industrial appliances is desirable. The interaction of calcite with biomolecules is also central to the study of sea creature shell formation and related species [1, 2]. Predictably $CaCO_3$ is the subject of extensive research, yet to our knowledge this important surface has not been addressed systematically by the technique of neutron reflection [3]. This technique can provide molecular level resolution of the surface structure normal to the interface and has been demonstrated to be a major asset in the study of adsorbed layers, most particularly at the air-liquid and solid–liquid interfaces [4]. In recent years many important systems have been studied using neutron reflection, however, in the study of solid–liquid interfaces, these measurements have been dominated by studies on silicon/silica [5–7] or alumina [8, 9] and rather few other solid substrates have been exploited.

The most stable polymorph of $CaCO_3$ under room temperature and pressure is calcite, shown in Fig. 1 [11]. Large, essentially perfect crystals with well defined crystal faces that are ideal for the reflection technique can be obtained of this polymorph. Here, we focus on the most stable (104) face, shown as the shaded area in Fig. 1.

$CaCO_3$ is slightly soluble in water with a solubility of 3.7×10^{-9} mol dm^{-3} under standard conditions [12, 13]. The dissolution of $CaCO_3$ is coupled to pH and prevalent carbon dioxide pressure (P_{CO2}) as listed in Table 1. Similar speciation reactions occur at the surface in contact with water. It can be shown that $CaCO_3$ dissolution increases with increasing P_{CO2} and/or decreasing pH. In the absence of added acid or base, solutions saturated with $CaCO_3$ reach a different equilibrium pH depending on the amount of CO_2 present. In the absence of CO_2 dissolution is minimal with an equilibrium pH of 9.91 whereas dissolution increases and the pH reaches 8.23 in equilibrium with atmospheric CO_2 (3.5×10^{-4} atm). The pH is therefore a key indicator of the equilibrium state of the system.

In order to control both surface speciation and surface roughness, a parameter linked to dissolution and of crucial importance in neutron reflection, P_{CO2} and pH need to be controlled carefully.

Neutron Reflection Theory

The details of the neutron reflection technique can be found elsewhere, e.g. [4]. Here the general approach is outlined. The geometry of a neutron reflection experiment is illustrated in Fig. 2. In specular reflection, of interest here, the incident angle θ_i is equal to the reflected angle θ_f and

S.M. Clarke (✉)
[1]BP Institute and Department of Chemistry, University of Cambridge, Madingley Rise, Madingley Road, Cambridge CB3 0EZ, UK
e-mail: stuart@bpi.cam.ac.uk
[2]BP Exploration Operating Company Ltd, Chertsey Road, Sunbury on Thames, TW16 7LN, UK
[3]Institut Laue-Langevin, 6 rue Jules Horowitz, BP 156, Grenoble Cedex 9, 38042, France
[4]ISIS, Rutherford Appleton Laboratory, Chilton, Didcot, OXON, UK

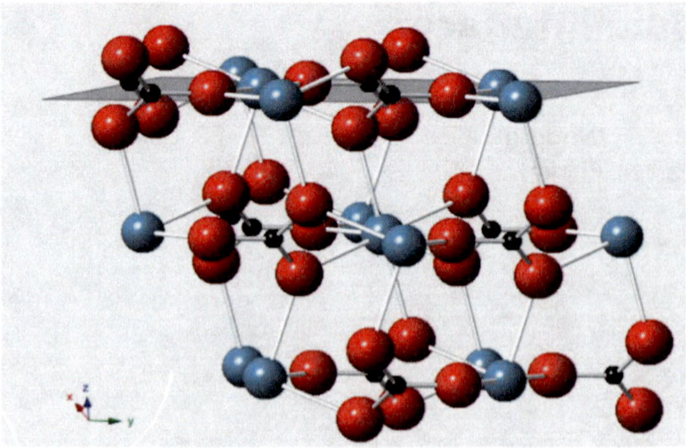

Fig. 1 Calcite cleavage rhombohedron; [10] oxygen-*red*, carbon-*black*, calcium- *blue* (images generated using CrystalMaker®)

Table 1 Solution reactions of calcium carbonate in water with respective equilibrium constants [12–14]

$CaCO_3 \rightleftharpoons Ca^{2+}+CO_3^{2-}$	$K_{sp}=3.7\times10^{-9}$ M
$H_2CO_3 \rightleftharpoons HCO_3-+H^+$	$K_{a1}=2.50\times10^{-4}$ M
$HCO_3^- \rightleftharpoons CO_3^{2-}+H^+$	$K_{a2}=5.61\times10^{-11}$ M
$CO_2(g) \rightleftharpoons CO_2(aq)$	$k_H=29.76$ atmM^{-1}
$CO_2(aq)+H_2O \rightleftharpoons H_2CO_3$	$K_h=1.7\times10^{-4}$ M^{-1}
$H_2O \rightleftharpoons H^+ + OH^-$	$K_W=1.0\times10^{-14}$ M

the scattering is elastic. Specular reflection provides insight into the structure normal to a surface. Off-specular reflection ($\theta_f \neq \theta_i$) can be used to extract in-plane structural information in some situations and will not be discussed further in this report [15].

The momentum transfer of specularly reflected neutrons, $\mathbf{q_z}$, is defined by

$$\mathbf{q_z} = \mathbf{k_f} - \mathbf{k_i}$$

$$|\mathbf{q}_z| = \frac{4\pi \sin\theta}{\lambda}$$

$\mathbf{k_i}$ and $\mathbf{k_f}$ are the wave vectors of incident and reflected neutrons respectively and λ is the wavelength. In a typical neutron reflection experiment, the variation of reflected intensity with $\mathbf{q_z}$ is measured. In a time-of-flight experiment, the angle of incidence is fixed and the variation in intensity with wavelength is considered. The different wavelengths are determined by the time the neutron takes to travel between a start point and the detector. Alternatively θ_i can be swept at constant λ, the so-called monochromatic mode.

Neutron reflection can be treated in an analogous way to the reflection of visible light. A neutron refractive index of non-absorbing materials can be defined as:

$$n = 1 - \frac{\lambda^2 \rho}{2\pi}$$

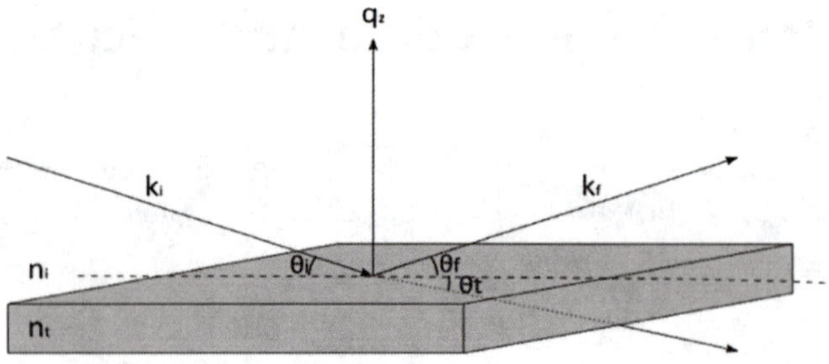

Fig. 2 Reflection geometry in a neutron reflection experiment; n_i, n_t refractive indices of bulk phases, k_i, k_f wave vectors of incident and reflected neutrons, θ_i, θ_f, θ_t angle of incidence, reflection and transmission respectively

where ρ is the scattering length density (SLD) of the material, essentially the variable that identifies how strongly a component scatters neutrons. It is defined as

$$\rho = \frac{\sum \boldsymbol{b}_{coh}}{v_M}$$

where v_M is the molecular volume and $\mathbf{b_{coh}}$ the coherent scattering length, a nuclear property that varies across the periodic table and even between isotopes. Typical values of ρ of relevance to this study are given in Table 2.

'Contrast variation' takes advantage of the variation of scattering length between isotopes of the same element [16]. Of particular importance in soft matter studies are the different neutron scattering abilities of hydrogen and deuterium. The scattering power of the solvent can be tuned by appropriate combinations of heavy and light solvent (e.g. H_2O and D_2O, D=^{2}H) while the chemical behaviour is essentially unaffected. For example, the contrast of hydrogenated solvent is particularly useful to highlight the calcite substrate roughness whereas contrast-matched solvent (a mixture of light and heavy solvent that matches the SLD of the substrate) allows identification of any adsorbed layers. The advantages of deuterated solvent are ease of alignment, short count times and the ability to probe both substrate roughness and adsorbed layers simultaneously. The combination of all three contrasts aids in providing a unique structural solution [16]. Note that in the presence of adsorbed organic molecules the possibilities of contrast variation increase substantially as various combinations of contrast-varied solvent and contrast-varied adsorbate become available.

Neutron reflectivity data analysis relies on fitting the calculated reflectivity profile from a model structure of the interface to the experimental data. The calculated reflectivity is compared to the experimental data and structural parameters varied until a best fit between calculated and experimental data is obtained.

In the optical matrix method the interface is divided into layers with defined thickness and composition [17]. Reflec-

Table 2 Scattering length densities of selected compounds

Material	$\rho(10^{-6}\ \text{Å}^{-2})$
H_2O	−0.56
D_2O	6.40
$CaCO_3$	4.70
d_8-toluene	5.66
d_4-methanol	5.80
d_{16}-heptane	6.30

tion is considered from each interface and the reflectivity determined from interference of all reflected signals. Each layer can be represented by a characteristic matrix:

$$M_i = \begin{bmatrix} \cos\beta_i & -\left(\frac{i}{p_j}\right)\sin\beta_i \\ -ip_j\sin\beta_i & \cos\beta_i \end{bmatrix}$$

where $p_j = n_j \sin\theta_j$ and β_i the phase shift neutrons undergo as they are transmitted and reflected in the layer. The matrix of the system is then given by the product of the characteristic matrices of all layers:

$$M = \prod_0^i M_i$$

Finally the reflectivity of such a system can be calculated from:

$$R = \left|\frac{(M_{11} + M_{12}p_s)p_a - (M_{21} + M_{22})p_s}{(M_{11} + M_{12}p_s)p_a + (M_{21} + M_{22})p_s}\right|^2$$

Experimental

Materials

Calcite Single Crystals

Calcite crystals of high optical clarity (Iceland Spar) were purchased from Moussa Direct Ltd. and cut by Crystran Ltd. The crystals were of either cuboidal or rhombohedral shape with sizes of $30 \times 30 \times 10\ \text{mm}^3$ to $40 \times 40 \times 10\ \text{mm}^3$.

Solvents

D_2O (98.5% deuteration) has been obtained from the ISIS and ILL deuteration facilities. d_8-toluene, d_4-methanol and d_{16}-heptane (all 99% deuteration) were purchased from Sigma-Aldrich and used without further purification.

Polishing

Calcite can readily be cleaved to expose the (104), (−114) and (0–14) faces. However cleavage is not suitable to obtain large atomically flat areas as required for neutron reflection. Thus polishing of the crystals is necessary but complicated by the relatively soft nature of calcite. Calcite has a Moh's hardness of 3, whereas, for comparison, silicon and sapphire have a hardness of 7 and 9 respectively [14]. Hence obtaining calcite surfaces suitable for neutron reflection poses a challenge. In order to achieve minimum roughness as required for neutron reflection, two polishing procedures and cleaving of the crystals were assessed. In polishing procedure A the crystals were smoothed using silicon carbide and finished with 1 μm diamond paste. In procedure B, undertaken by Crystran Ltd., the crystals were smoothed with aluminium oxide and finished with colloidal silica [18]. The surface roughness was determined with profilometry and atomic force microscopy (AFM), as discussed below. Prior to running a crystal on the neutron source it is also convenient to perform an initial X-ray reflection experiment to confirm the surface roughness. In this way only the best crystals can be prepared for the neutron experiment. All crystals finally used for neutron reflection experiments were polished by Crystran Ltd. using procedure B.

Sample Cleaning

The solubility of $CaCO_3$ discussed above poses the difficulty of cleaning the sample of surface impurities, as most conventional cleaning methods prior to a neutron reflection experiment require strong acid (e.g. concentrated nitric acid or piranha solution) which would badly roughen the calcite surface or even dissolve the crystal. Instead, calcite crystals were cleaned by UV/ozone treatment using a Novascan PSD-UV4 and plasma treatment using a Diener FEMTO plasma system. The effectiveness of the two methods was assessed by X-ray photoelectron spectroscopy (XPS). As discussed below, UV/ozone treatment for 10 min was the most effective method and has been applied for all future measurements immediately before mounting of the crystals in the sample cell [3].

Control of Surface Dissolution

Even slight dissolution of the sparingly soluble calcite can have a detectable effect on surface roughness. To control dissolution, water pre-saturated with $CaCO_3$ was used. This has been prepared stirring powdered $CaCO_3$ (Sigma-Aldrich, ≥99%) with H_2O and D_2O in a closed bottle for 1 day. The final pH/pD was around 9.9 as expected for a

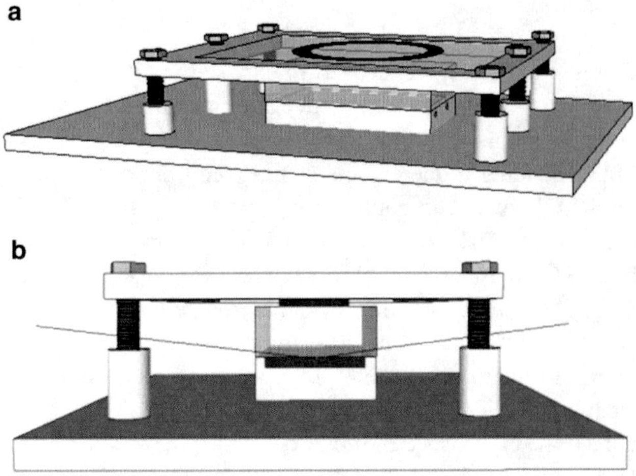

Fig. 3 (**a**) Illustration of a single cell and its mounting arrangements and (**b**) cross-section through a cell illustrating the calcite crystal, the PTFE trough and liquid showing the neutron beam path

saturated solution in a closed system (no exchange with atmospheric CO_2). Total organic carbon analysis revealed no organic carbon present in water saturated with the $CaCO_3$ used (TOC = -0.03 ± 0.07 ppm). Water contrast-matched to calcite (CMW) contained a mixture of D_2O and H_2O in the ratio 0.765:0.235 v/v. Saturated water will be referred to as H_2O(sat.), D_2O(sat.) and CMW(sat.) respectively.

Sample Cells

Design: The sample cell was designed specifically for this experiment to create a solid–liquid interface for neutron reflection. A key issue is to provide a good seal to obtain an effective solid–liquid contact and contain the liquid even in a vertical alignment without fracture of the reasonably fragile calcite crystals. The calcite crystals were clamped between a PTFE trough and an o-ring set into a clear Perspex lid by three bolts, as illustrated in Fig. 3 below. The edges of the PTFE trough were chamfered to improve the seal for the larger crystals by an increase in the local pressure. Three independently adjustable spacers were used to ensure an even pressure. The o-ring was visible through the clear plastic lid and used to help ensure even compression of the cell. This was particularly important with the larger calcite crystals. Later modifications of the cell used a square ring so that the clamping pressure could be applied more neatly over the edges of the crystal, rather than at the centre of the crystal which tended to crack the crystals. The liquid cell volume was minimised to reduce use of deuterated solvents and varied from 2.7 ml for the smallest crystals to 4.8 ml for the largest crystals.

Cleaning: All PTFE troughs and glassware were cleaned in concentrated HNO_3 (Sigma-Aldrich, $\geq$69%) for 6–12 h and washed with copious amounts of ultra pure water. Steel cells, plastic lids and spatulas were cleaned in Decon 90 for 6–12 h before copious washing with ultra pure water.

Reflection Measurements

Neutron Reflection measurements were performed at the instruments CRISP at the Rutherford Appleton Laboratory, U.K., and at the instrument D17 at the Institut Laue-Langevin, France. Detailed description of the instruments can be found in references [19, 20]. The sample geometry at CRISP was horizontal and vertical at D17. Where selected samples were repeated, excellent agreement was found between the two devices.

All samples were aligned using the highly scattering contrast of deuterated solvent. The vertical set-up at D17 allowed in-situ sample changes between different contrasts. At CRISP the sample holder had to be removed from the goniometer for sample changes. However, it was found that the alignment was still highly repeatable. This proved to be particularly useful for the contrast of contrast-matched water since alignment was complicated by the low reflected intensity. We are hence moderately confident that the very weak reflection from this contrast is a genuine indication of the absence of significant adsorbed material. During sample changes, the cell was rinsed three times with the subsequent solution to ensure complete exchange of the liquid.

Reflectivity was recorded in time-of-flight mode on both devices. Data was normalised by measuring the direct beam so that the reflectively can be calculated directly. The beam resolution ($\Delta\lambda//\lambda$) on CRISP was 4.06% for all $\mathbf{q_z}$. On D17, it was 1% for $\mathbf{q_z} < 0.02$ Å^{-1} and 4% for $\mathbf{q_z} > 0.02$ Å^{-1}. At D17 the background was subtracted by taking into account scattering recorded on either side of the specular reflected signal on a two-dimensional detector. At CRISP no background was subtracted but was included in the data analysis routines. The data was analyzed using the software RasCAL.

Results and Discussion

Preliminary Investigations

Sample Size

Neutron reflection requires a reasonably large surface area (ideally at least 4 × 4 cm) to minimize count times and maximize the signal-to-noise ratio. The sample size is limited, however, by the transmission of the sample to the neutrons.

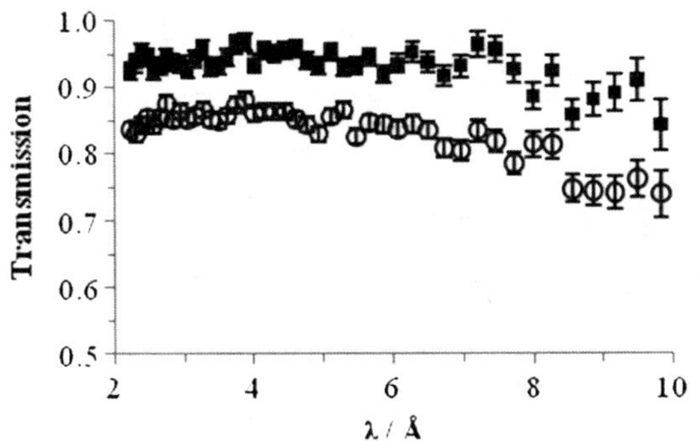

Fig. 4 Transmission of neutrons through (■) 12 mm and (○) 30 mm thick calcite crystal as a function of neutron wavelength

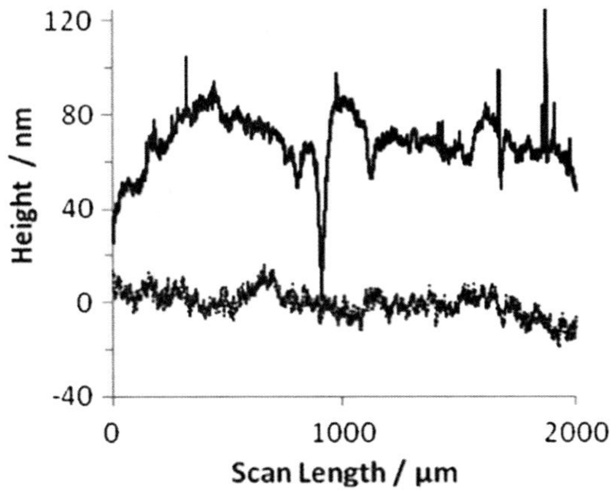

Fig. 5 Profilometry results of polished crystals: *solid line* finished with diamond paste (shifted by 70 nm for clarity); *dotted line* finished with colloidal silica

For a calcite crystal of 30 mm length the transmission was found to be 87% (Fig. 4). Crystals of path length 30–40 mm are therefore considered to be ideal for neutron reflection experiments.

Roughness

Neutron reflection requires smooth surfaces for maximum resolution. Cleaving of the crystals produced locally atomically flat areas (data not shown) but separated by visible cleavage steps. While a cleaved surface would be ideal for reflection owing to cleanliness and roughness, it was impossible to seal cells containing crystals with cleavage steps and an alternative polishing procedure was established, as described above. Figure 5 shows profilometry height scans of polished surfaces over a length of 2 mm. The root-mean square roughness (RMS) is defined as

$$RMS = \sqrt{\frac{1}{n}\sum_{i=1}^{n} y_i^2}$$

where n is the number of data points and y_i the height at distance i. The RMS over a length of 2 mm in polishing procedure A was 12.5 nm (solid line in Fig. 5), unacceptably large for neutron reflection. Polishing procedure B (dotted line in Fig. 5) produced a RMS of 6.2 nm. Since profilometry is only a crude measure this representative crystal was further investigated by AFM, see Fig. 6, and a RMS of 30 Å could be confirmed. This roughness is still high relative to silicon or alumina, reducing the useful q-range to $q < 0.2$ Å^{-1}. Yet it is sufficiently low to characterise the bare calcite-liquid interface as well observe adsorption of sufficiently thick layers. The polishing procedure has since been improved further and neutron reflection proved roughnesses in the range 7–45 Å with generally lower values in later experiments.

Evaluation of Surface Cleanliness

Polished calcite crystals were cleaned by plasma and UV/ozone treatment respectively and the thus achieved cleanliness compared to a freshly cleaved surface assessed using XPS. The XPS data is shown in Table 3 and indicate that the as received crystal contained a significant amount of impurities. However, both the Plasma and UV treated samples show reductions in the impurity levels of adventitious carbon. Even the sample freshly cleaved in air shows evidence of some impurities. The XPS data suggest that plasma treatment was less successful in removing adventitious carbon. Surprisingly, it also appeared to give rise to the presence of aluminium and fluorine. Although their origin is not well understood it is likely that aluminium will originate from contamination in the plasma cleaner. Fluorine may be present as trace element in the calcite. The 10 min UV/ozone treatment generated very similar sample cleanliness as the cleaved surface. Extending the exposure to 45 min had an adverse effect which is thought to arise from the generation of an activated surface that is more reactive toward atmospheric organic impurities. Hence we conclude that 10 min UV/ozone treatment is appropriate for cleaning calcite surfaces in preparation of a neutron reflection experiment.

Calcite-Water Interface

The reflectivity of a bare crystal in the two contrasts of H_2O (sat.) and D_2O(sat.) are shown in Fig. 7. The profiles were fitted with the scattering length density parameters expected for calcite, H_2O and D_2O (see Table 2) and very reasonable agreement was found between the expected and experimental data. A roughness of 22 Å was included in the

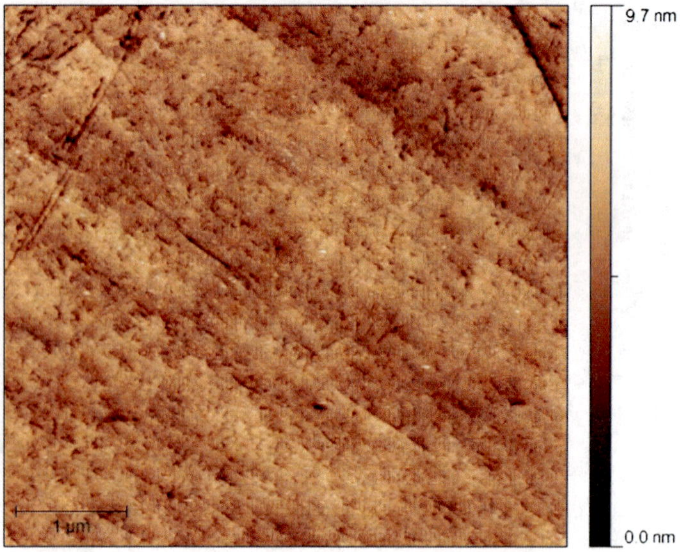

Fig. 6 AFM image of polished calcite surface, RMS roughness 30 Å

Table 3 XPS analysis of calcite surface after various cleaning procedures, all values in atom-percent

	Ca 2p	O 1s	C 1s (carbonate)	C 1s (adventitious)	Al 2p	F 1s	N 1s	Na 1s	Fe 2p
As received	12.7	43.7	15.6	28.0	0.0	0	0	0	0
Cleaved	16.1	54.5	19.6	8.8	0.0	1.0	0	0	0
Plasma, 10 min	7.4	56.7	7.3	12.9	11.9	2.0	1.6	0	0.2
Plasma, 45 min	0.6	57.8	0.0	14.2	23.7	1.9	1.6	0	0.2
UV/ozone, 10 min	15.3	55.2	19.7	8.6	0.0	0.8	0	0.4	0
UV/Ozone, 45 min	13.9	52.0	18.0	14.5	0.0	1.1	0	0.5	0

model. There was no evidence of a hydrocarbon layer at the interface in the contrasts of H_2O(sat.) and D_2O(sat.) and essentially no reflectivity was observed in contrast-matched water. Note that the rough interface could equally be represented by a calcite substrate with 3 Å roughness and a 30 Å thick layer consisting of 60% calcite and 40% water, We conclude that this data shows that it is possible to achieve neutron reflection from the calcite-water interface.

Several other $CaCO_3$ crystals have also been investigated with these contrasts. In general, their reflectivity is very similar to that given in Fig. 7 differing predominantly in the level of roughness, which can be as low as 7 Å and as large as 45 Å.

Calcite in contact with saturated water is not expected to dissolve as the water phase already contains equilibrium concentrations of all species mentioned in Table 1, although there should be a dynamic equilibrium at the surface. Indeed, Fig. 8 illustrates that there is no change in reflectivity between scans immediately after exposure to saturated water and after 20 h exposure. Hence we conclude that there is minimal surface dissolution under these conditions. Both reflectivity profiles of this crystal could be fitted with the expected scattering length densities of calcite and water. In this case the fitted roughness was 45 Å.

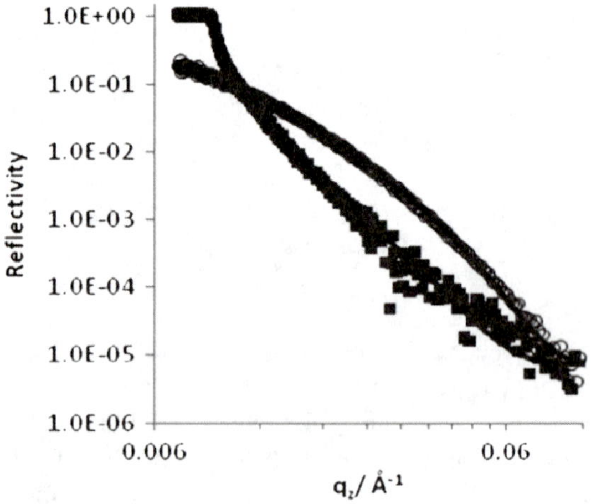

Fig. 7 Neutron reflectivity profiles of bare crystal in H_2O (○) and D_2O (■). *Solid lines* are fits based on the expected scattering length densities of calcite and water, with a roughness of 22 Å

Surface 'healing' of a cleaved calcite surface in contact with super-saturated water had been observed by X-ray reflectivity [21]. While the experimental conditions have been repeated here (data not shown) no evidence of signifi-

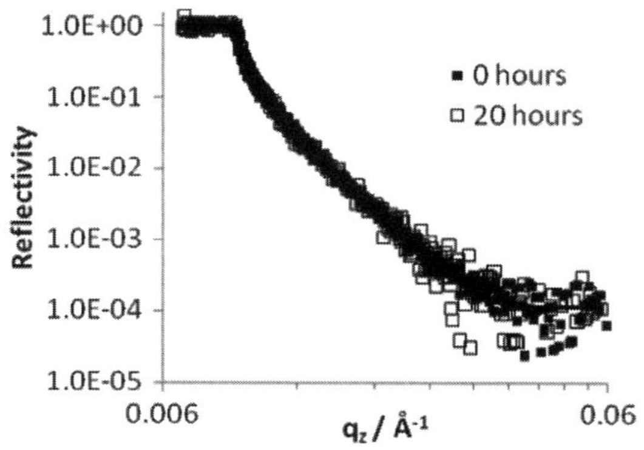

Fig. 8 Neutron reflectivity profile of calcite-D2O(sat.) interface at 0 h (■) and 20 h (□) contact time

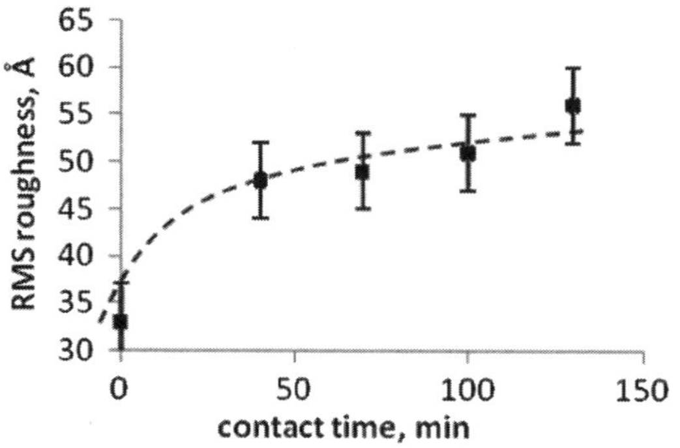

Fig. 9 Increase of RMS roughness with time, for a calcite crystal in contact with water at pH 6.5. The *dotted line* is a guide to the eye

Table 4 RMS roughness as function of pH and contact time

Water conditions	Contact time/min	Substrate roughness/Å
Saturated water	100	10±4
Pure water (pH 6.5)	100	33±5
Acidified water (pH 4.8)	40	48±4
	70	49±4
	100	51±4
	130	56±4

cant reductions in roughness was observed. It was previously shown that cleaved and polished calcite surfaces can behave quite differently: in a X-ray reflection study of the calcite-air interface Bohr et al. demonstrated that a water film deposited from vapour was twice as thick on the rougher polished surface [18]. This may explain why no surface 'healing' was observed on the polished crystals whereas it has been reported on cleaved samples [21].

The equilibria in Table 1 suggest that little dissolution should occur in water with basic pH. Indeed, no increase in surface roughness has been observed in the neutron reflection data after exposure to water of pH 10.2 without added $CaCO_3$ (data not shown). We conclude that the calcite surface is stable under conditions predicted from the bulk solution phase equilibria.

In contact with pure water (pH 6.5) for 2 h, the surface roughness was found to increase from 10 to 33 Å (data not shown). From the bulk phase equilibria (Table 1) dissolution of 1.2×10^{-4} mol/L $CaCO_3$ is required for equilibrium in absence of CO_2 (the situation in the closed neutron reflection cell). Previous AFM studies showed that calcite dissolution proceeds via etch pitch formation at localised defect sites rather than dissolution of a uniform layer [22, 23]. An increase in roughness, as observed here, is therefore considered appropriate. The experimental finding agrees particularly well with a previous AFM study of the dissolution of polished calcite in pure water [22]. The authors report a linear increase in surface roughness during the first 60 min of exposure, increasing from about 10 to 30 Å [22]. The roughness then stayed roughly constant before decreasing slightly. This behaviour was attributed to the surface roughness increasing up to a point where the entire surface was covered in etch-pits so that further dissolution caused the roughness to stay constant or decrease if etch-pits started to merge [22]. However this reduction was only of the order of a few Å and may be within the error of the experiment and was not observed in our study. Specular neutron reflection has very poor lateral resolution but very high vertical resolution. The excellent agreement between the two studies provides confidence in the model and the validity of the technique of neutron reflection from the calcite-water interface [22].

Lastly, the behaviour of the calcite surface under acidic conditions (pH 4.8) has been investigated (reflectivity data not shown). Here, the roughness increased significantly within the first 40 min contact time and continued to rise more slowly until the experiment was terminated, as illustrated in Fig. 9 and Table 4. The change in dissolution rate corresponds well to that observed by Bisschop et al., as discussed above, albeit under slightly more acidic conditions [22].

Calcite-Oil Interface

The interface between calcite and hydrophobic solvents has also been investigated. While surface dissolution is not a problem in contact with organic solvents, the volatility and viscosity of the solvents poses additional technical challenges. The hydrophobic character of the PTFE trough means that cells containing water could be sealed by applying only moderate pressure. It was found that significantly higher pressure was required to seal cells containing hydrocarbons, increasing the risk of fracture of the fragile crystals. To circumvent this problem, the PTFE troughs were chamfered to achieve a locally higher pressure. In addition, the o-ring in the initial set-up was exchanged by a square ring as the compression base so that the clamping pressure could be

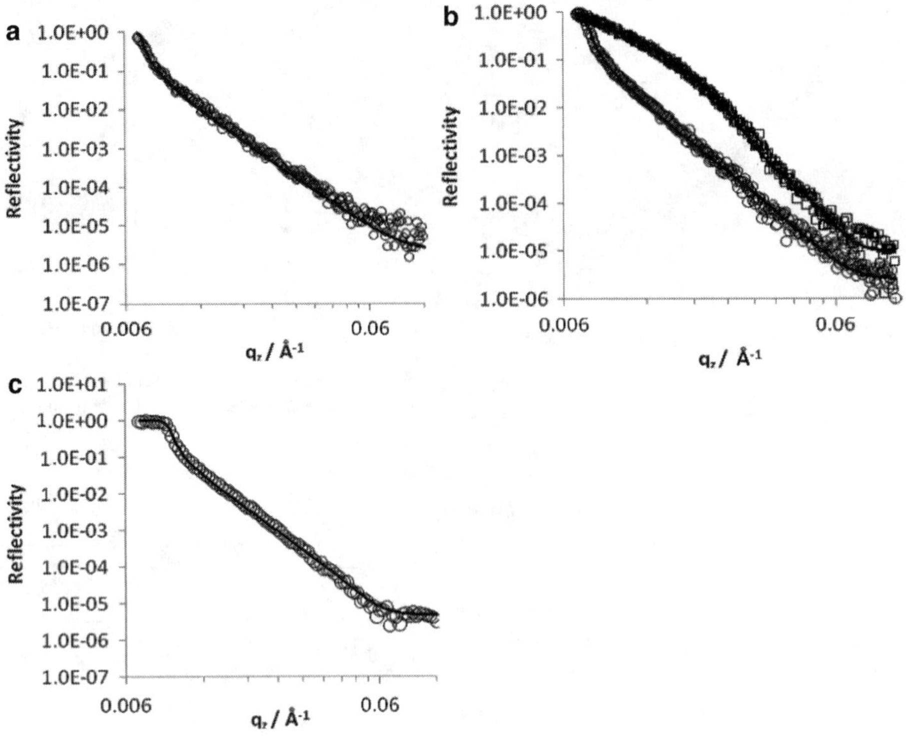

Fig. 10 Neutron reflectivity profiles of (**a**) calcite #1/d_8-toluene; (**b**) calcite #3/d_4-methanol (○) and calcite #3/air/d_4-methanol after cell leak (□); (**c**) calcite #2/d_{16}-heptane; calculated reflectivity shown as *solid lines*

Table 5 Fitting parameters

Solvent	SLD/10^{-6} Å^{-2}	Substrate roughness/Å	Thickness of air layer/Å	Roughness of air layer/Å
d_8-toluene	5.60±0.03	10±2	n.a.	n.a.
d_{16}-heptane	6.23±0.03	19±2	n.a.	n.a.
d_4-methanol	5.74±0.04	20±2	n.a.	n.a.
d_4-methanol (after leak)	5.74±0.03	23±3	150 ±10	58±5

applied more neatly over the edges of the crystal and PTFE trough. Using these measures, it was possible to record reflectivity curves from the interface of calcite with hydrocarbon solvents, as illustrated in Fig. 10. The corresponding fitting parameters are summarised in Table 5. In all cases the calculated scattering length densities are as expected from the literature. Figure 10c illustrates the effect of a cell leak in a horizontal sample alignment. The squares were recorded after some solvent has leaked and was replaced by a thin layer of air which caused the reflectivity to lift in comparison to a sealed sample cell. Increasing the clamping pressure was sufficient to seal the sample and a reflectivity as expected from d_4-methanol could be recorded (circles). Since the scattering length densities of air and hydrogenous material are similar (close to or equal to zero), leaking of the cell and displacement of liquid with air has the same effect as adsorption of a hydrogenous layer. In the study of adsorption at the solid–liquid interface leaking of the cell has hence to be eliminated. The sample cells were designed such that the clamping pressure required to seal the cells effectively can be set at the beginning of the experiment. The sample is then run in absence of adsorbing species to verify that no leaking occurs at the set clamping pressure so that later changes in reflectivity is presence of solutes can reliably be attributed to adsorption of hydrogenous material.

Conclusions

Here we have presented the neutron reflection from the calcite/water interface. The experimental data are shown to be in good agreement with that expected from the compositions of the materials present suggesting that the measurements are a good representation of the system. In addition we have been able to address the key issues that enable these measurements to be made such as crystal size, cleanliness and roughness: The CMW contrast is a sensitive measure of impurities at the interface and suggest good cleanliness. The roughness is sufficiently small so that adsorbed layers can be identified and characterised. Further work is underway to improve this aspect and enhance the accessible q-range and resolution of the measurements.

Acknowledgements We thank BP and the Cambridge European Trust for financial support for this work and the ILL and ISIS staff for allocation of beam time (ISIS: RB920097, RB920098; ILL: 9-12-257, 9-11-1461) and technical assistance with neutron reflection measurements. We are grateful to Giovanna Fragneto, Adrian Rennie, Tommy Nylander and Arwel Hughes for help with data analysis and the EPSRC and Emily Smith for XPS measurements (University of Nottingham, Open Access Scheme funded by EPSRC).

References

1. Sand KK, Yang M, Makovicky E, Cooke DJ, Hassenkam T, Bechgaard K, Stipp SLS (2010) Langmuir 26:15239
2. Meldrum FC (2003) Int Mater Rev 48:187
3. Stocker IN, Lee SY, Miller KL, Collins IR, Webb KJ, Wildes A, Kinane C, Clarke SM (2011) 16th European symposium on improved oil recovery, Cambridge, UK
4. Penfold J, Thomas RK (1990) J Phys Condens Matter 2:1369
5. Li ZX, Weller A, Thomas RK, Rennie AR, Webster JRP, Penfold J, Heenan RK, Cubitt R (1999) J Phys Chem B 103:10800
6. Li ZX, Lu JR, Fragneto G, Thomas RK, Binks BP, Fletcher PDI, Penfold J (1998) Colloid Surf A Physicochem Eng Asp 135:277
7. Fragneto-Cusani G (2001) J Phys Condens Matter 13:4973
8. Hellsing MS, Rennie AR, Hughes AV (2010) Langmuir 26:14567
9. Hellsing MS, Rennie AR, Hughes AV (2011) Langmuir 27:4669
10. Markgraf SA, Reeder RJ (1985) Am Mineral 70:590–600
11. Reeder RJ (1983) Carbonates: mineralogy and chemistry. Mineralogical Society of America, Washington, DC
12. Butler JN (1982) Carbon dioxide equilibria and their applications. Addison-Wesley, Reading
13. Stumm W, Morgan JJ (1996) Aquatic chemistry: chemical equilibria and rates in natural waters. Wiley-Interscience, New York
14. Lide DR (2000) CRC handbook of chemistry and physics: a ready-reference book of chemical and physical data. CRC Press, Boca Raton
15. Li ZX, Lu JR, Thomas RK, Penfold J (1996) Faraday Discuss 104:127
16. Crowley TL, Lee EM, Simister EA, Thomas RK (1991) Physica B: Condens Matter 173:143
17. Heavens OS (1955) Optical properties of thin solid films. Butterworths Scientific Publications, London
18. Bohr J, Wogelius RA, Morris PM, Stipp SLS (2010) Geochim Cosmochim Acta 74:5985
19. Penfold J (1991) Physica B: Condens Matter 173:1
20. Cubitt R, Fragneto G (2002) Appl Phys A: Mater Sci Process 74:s329
21. Chiarello RP, Sturchio NC (1995) Geochim Cosmochim Acta 59:4557
22. Bisschop J, Dysthe DK, Putnis CV, Jamtveit B (2006) Geochim Cosmochim Acta 70:1728
23. Hillner PE, Manne S, Gratz AJ, Hansma PK (1992) Ultramicroscopy 42–44:1387

Progr Colloid Polym Sci (2012) 139
DOI: 10.1007/978-3-642-28974-3_17

Passive Microrheology for Measurement of the Concentrated Dispersions Stability

Christelle Tisserand, Mathias Fleury, Laurent Brunel, Pascal Bru, and Gérard Meunier

Abstract This work presents a new technique of passive microrheology for the study of the microstructural properties of soft materials such as emulsions or suspensions. This technology uses Multi Speckle Diffusing Wave Spectroscopy (MS-DWS) set-up in a backscattering configuration with video camera detection. It measures the mean displacement of the microstructure particles in a spatial range between 0.1 and 100 nm and a time scale between 10^{-2} and 10^{5} s. Different parameters can be measured or obtained directly from the Mean Square Displacement (MSD) curve, including a fluidity index, a solid–liquid balance, an elasticity index, a viscosity index, a relaxation time and a MSD slope.

Using this technique the evolution of the microstructure, the restructuring after shearing and the variation of the viscoelastic properties with temperature or pH can be measured allowing the physical stability of emulsions or suspensions to be forecasted.

This work focuses on the evolution of the viscoelastic properties of emulsions and suspensions to follow their stability over time and shows the advantages of using a non-invasive method to detect nascent destabilisation of the microstructure.

Introduction

For many years, soluble polymers have been routinely added to emulsions and suspensions in order to improve their stability in respect of creaming or sedimentation. These polymers give these concentrated dispersions gel-like behaviour, leading to viscoelastic properties. Formulators need to characterize these new products as they have new rheological properties such as consistency, spreadability, shape stability, workability or physical stability. Thus it is crucial to characterize the rheological behaviour using properly adapted techniques. Microrheology is a new domain of rheological methods studying viscoelastic behaviour of soft matter such as suspensions, emulsions, gels, colloidal dispersions at the micron length scale.

The optical method used in passive microrheology involves measuring the mean displacement of constituent particles (or droplets, fibres or crystallites contained in the material) as a function of time. This raw data provides information on the viscoelasticity of the sample when analysed appropriately. This technique enables the analysis of a product at rest (with zero shear), it is a non contact measurement (the product is not denaturated), the sample being monitored versus ageing time.

C. Tisserand (✉)
Formulaction, 10 Impasse Borde Basse, L'Union, 31240, France
e-mail: tisserand@formulaction.com

Experimental Set-up

Measurements

Measurements were performed using a Rheolaser Lab Diffusing Wave Spectroscopy instrument (Formulaction, France). It consists of Dynamic Light Scattering (DLS) extended to an opaque media. DLS is a well known method monitoring Brownian motion of particles in a diluted media in order to determine the particle size. In a DWS experiment (more precisely Multi Speckle-DWS in this case) [1, 2], a fixed coherent laser beam is applied 170 mm from the sample containing scaterrers (particles, droplets, fibers…). The sample is contained in a cylindrical glass cell of 20 mL with a 25 mm diameter. The light is multiply-scattered many times, by the particles into the sample, which leads to interfering backscattering waves (Fig. 1). An interference image called a "Speckle image" is detected by a multi-pixel detector. In dynamic mode, the scatterers motion (resulting from thermal energy k_bT with k_b being the Boltzman constant, and T the temperature) induces spot movements of the speckle image [3]. A patented algorithm [4]

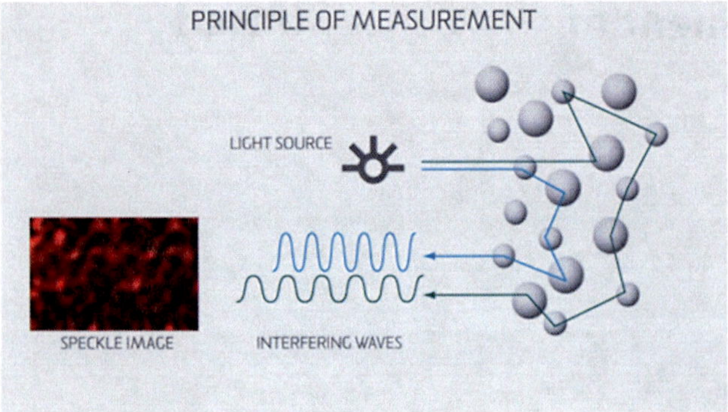

Fig. 1 Measurement principle (Multi-Speckle Diffusing Wave Spectroscopy)

enables the treatment of this speckle image in order to determine the scatterers mobility in terms of speed and displacement which are directly related to the samples viscoelastic properties. The MS-DWS technique enables the measurement of the viscoelasticity of samples by the microrheology method presented in the following chapter.

Experimental Systems

Emulsion

Oil-in-water emulsions were prepared at room temperature. The initial step consists in preparing a crude polydisperse emulsion, the so-called premixed emulsion. It is obtained by incorporating the dispersed phase (vegetal oil) under gentle manual mixing into a mixture of water, salt, surfactant (Tween 20) and preservative (potassium sorbate). The premixed emulsion is comprised of 50% oil, 2% surfactant, 1% NaCl in volume. An Ultraturrax (rotor stator mixer, speed 24,000 min^{-1}) was used to fragment the premixed emulsion to reduce the average diameter and obtain a lower degree of polydispersity. The average droplet size is 2 μm. The next step consists in preparing xanthan solution at 1% w/w. Sodium chloride is added to help the polymer to extend the chains. Then the emulsions are diluted with the xanthan solution to adapt a final volume fraction of 20% for the emulsion and variable concentrations for the xanthan polymer (0.12%, 0.15%, 0.25%, 0.40%). The mixture is then immediately sampled in a 20 mL glass cell and introduced into the instrument for analysis.

Paint

Commercial paints were analysed, described by the supplier as a classic one and a non-drip one (Tollens). The non drip one is presented as a paint containing higher amount of polymer, improving the elasticity of the system.

Microrheology Theory

Microrheology [5] consists of using micron sized particles to measure the local deformation of a sample resulting from an applied stress or thermal energy (~k_bT). When microrheology measurements are performed using an applied stress to displace the particles (optical tweezers, magnetic field), the method is called active microrheology. When microrheology measurements are performed by measuring the displacement of particles due to the thermal energy, that is to say the Brownian motion, the method is called passive microrheology [6].

The instrument used to perform this work uses the passive approach. The measurement is done at rest as no mechanical or external stress is applied. The unique force used to displace the particles is thermal energy which may be 10^{12} times lower than macroscopic mechanical stress provided by a rheometer.

The treatment of the speckle pattern with a patented algorithm enables the plotting of the Mean Square Displacement (MSD) of the particles versus decorrelation time. Decorrelation time is the observation time scale: initially (short time scale) the particle movement probes the solid component of the sample (elasticity) and then (longer time scales) the liquid component (Fig. 2).

There are two types of motion, Brownian and Ballistic, Brownian motion is diffusive which means that the particles do not move in any particular direction but randomly move everywhere within the available space. Thus the displacement is measured as a squared distance (in nm^2). It is a "mean" displacement because the displacement of many particles is tracked simultaneously with only one measurement.

Figure 3 gives the typical shape of the MSD for a purely viscous product: the MSD grows linearly with decorrelation time as the particles are completely free to move in the sample. This figure gives also the typical shape for the MSD of a viscoelastic product (concentrated emulsion, polymer solution with particles). Over very short observation times, the particles are free to move in the continuous phase. They are then blocked by their neighbours (or by polymers), and the MSD reaches a plateau. This is characteristic of the elasticity of the product, the lower the plateau, the tighter the network, and the stronger the elasticity. Then, at longer time scales, the scatterers are able to find a way to escape from the "cages" formed by neighbouring particles or polymers and the MSD grows as it would for a viscous fluid. This is characteristic of the macroscopic viscosity, as it corresponds to the speed of the particles in the sample.

In summary, the MSD is the viscoelastic fingerprint of the analysed product (Fig. 4):

- The lower the elastic plateau, the stronger the elasticity, the Elasticity Index corresponds to the inverse of the MSD plateau value;

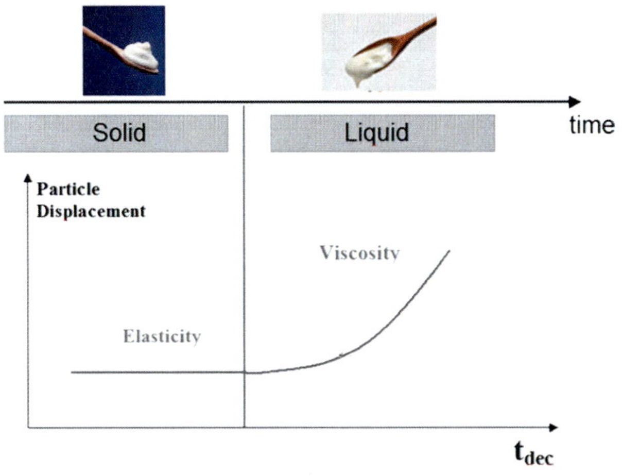

Fig. 2 Viscoelastic behaviour over time

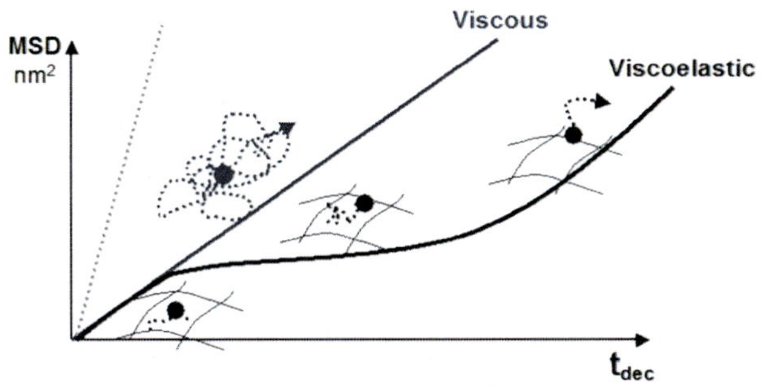

Fig. 3 Mean Square Displacement (MSD) of purely viscous and viscoelastic materials

- The Solid–liquid Balance (SLB) corresponds to the MSD slope at short decorrelation time: SLB = 0.5 means that the liquid and solid parts are equal, 0.5 < SLB < 1 means that the liquid behaviour dominates, 0 < SLB < 0.5 means that the solid behaviour dominates (gel behaviour).
- The longer the time needed by the particles to travel a distance, means they have a lower particle speed and a higher macroscopic viscosity, the Macroscopic Viscosity Index (MVI) corresponds to the inverse of the slope of MSD in linear scale. This MVI is linked to the well known macroscopic viscosity in Pa.s.

From the MSD one can measure:

- The viscosity and elasticity indexes which are a simple way of comparing the viscous and elastic behaviour of similar products,
- Parameters such as relaxation time or macroscopic viscosity,
- The elastic and viscous moduli G′ and G″. These can be calculated using the Generalized Stokes Einstein relation [7] by knowing precisely the particle size and assuming the sample is homogeneous.

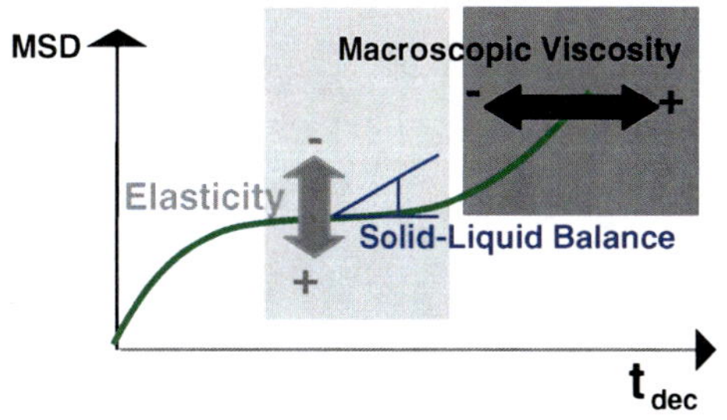

Fig. 4 Mean Square Displacement (MSD) as a viscoelastic fingerprint

Results

Emulsion Stability

Emulsions are commonly used in food and cosmetic industries. Their shelf life is very important for the supplier and also for the customer. Polymers are used to stabilize emulsions and provide them particular viscoelastic properties. These properties drive several end use properties such as physical stability or efficiency during use.

In this example, four emulsions with xanthan polymer are analysed and the goal is to determine their physical stability to rank them depending on this criteria. Xanthan is known as a non-absorbing polymer, will play the rolc of a dcpletion flocculent and create a transient gel with the emulsion [8]. The integrity of the gel persists for a finite period of time, then the structure collapses suddenly, local fractures appear, a percolation network forms then a denser creamed phase forms. This phenomenon is called delayed creaming [9].

Figure 5 gives the MSD curves at different ageing times for the sample with 0.12% xanthan. The measurements were started just after sample preparation.

The MSD curves first move from long to short displacement meaning an increase in the elasticity and from short to long decorrelation time (meaning an increase in the viscosity) (**step 1**). This step monitors the network flocculation and then stabilisation. Then at one moment, the curves move towards the left, linked to the sample destabilisation, the structure is breaking (**step 2**). This is the beginning of microstructure evolution.

The macroscopic viscosity index is directly obtained from the MSD curves (Fig. 6). Its evolution shows a first increase in step 1, then a drop corresponding to step 2. So when the viscosity begins to decrease, the sample becomes unstable at the microscopic scale. The drop of the viscosity index appears later when the xanthan concentration increases (Table 1). The more thickener is added to the sample, the more the creaming is delayed.

Table 1 gives also the destabilisation times corresponding to macroscopic destabilisation (eye observation). The micro-

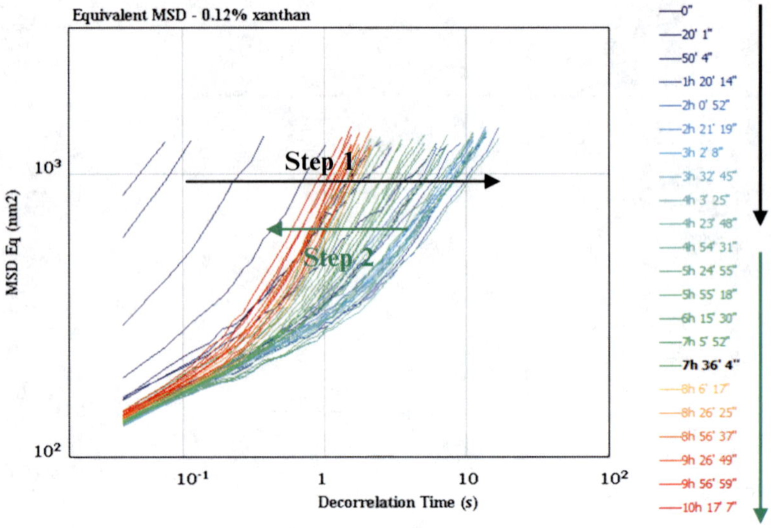

Fig. 5 MSD curves at different ageing times for an emulsion with 0.12% xanthan

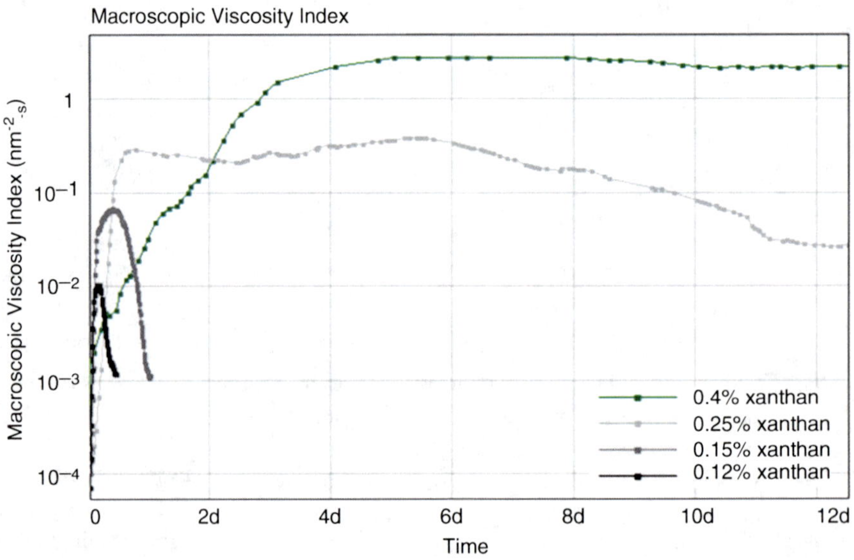

Fig. 6 Macroscopic viscosity index versus ageing time for all emulsions

scopic scale detects destabilisation phenomenon much earlier meaning that the microstructure begins to destabilize before the sample changes at the macroscopic level. For both methods, the more structured is the sample (the higher is the xanthan concentration), the longer it is stable (the longer is the viscosity drop time).

Paint Stability

Paints are mainly dispersions of pigments and binders stabilized with a polymer. Depending on the formula and polymer nature, the paint will be more or less stable and a transparent layer will or will not appear on top when the user opens the container.

The slope of the MSD at long decorrelation time provides information on the nature of the motion: if the slope is equal

Table 1 Stability time for all samples measured with microrheology (Extracted from Fig. 6)

Xanthan concentration in the emulsion (%)	Stability at the microscopic scale (time before the MVI drop) (h)	Macroscopic stability (eye observation) (h)
0.12	4.25	65
0.15	11	168
0.25	132	>864
0.40	1,008	>1,536

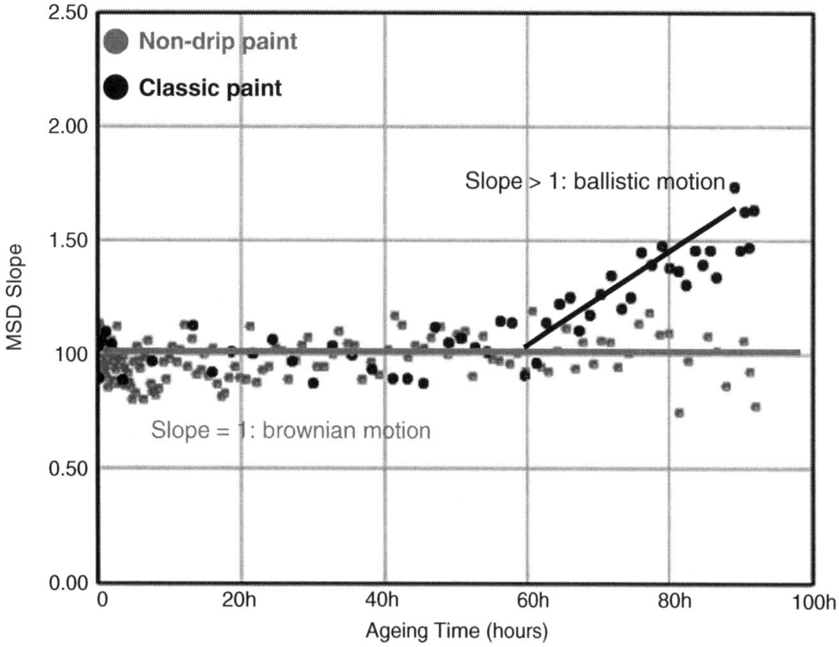

Fig. 7 MSD slope measured at long decorrelation time versus ageing time

to 1, the motion is brownian and if the slope is greater than 1, the motion is ballistic. A ballistic motion corresponds to a specific motion of the particles, which is not purely brownian. Very often, the ballistic motion corresponds to sedimentation or creaming phenomenon.

By examining the MSD slope versus ageing time, it is possible to monitor stability. Figure 7 monitors the MSD slope versus ageing time for both paints.

For the non-drip paint, the slope remains equal to 1 for 100 h, meaning that the particle motion is purely brownian. The particles are completely stable in the paint. For the classic paint, the slope increases after 60 h, meaning that ballistic motion is occurring in the sample. The pigment particles are settling toward the bottom, thus indicating the initial time of destabilisation of the paint.

Conclusion

The microrheology method using MS-DWS enables the viscoelasticity of soft matter to be characterized in real time and linked to the desired end user properties. The measurement is performed using a non-contact method which enables sample analysis without the application of an external stress. It has been demonstrated that the method has promise as a way of monitoring physical stability. We have seen that it can be applied to both emulsions and suspensions.

References

1. Weitz DA, Pine DJ (1993) In: Brown W (ed) Dynamic light scattering, the method and some applications. Oxford University Press, Oxford, 652
2. Weitz DA, Pine DJ (1993) In: Brown W (ed) Dynamic light scattering. Oxford University Press, New York
3. Brunel L, Snabre P (2003) EP 1,664,744
4. Brunel L (2009) Formulaction, WO2010130766
5. Gardel ML, Valentine MT, Weitz DA (2005) 1 Microrheology. Department of Physics and Division of Engineering and Applied Sciences, Harvard University, Cambridge, MA 02138
6. Bellour M, Skouri M, Munch J-P, Hébraud P (2002) Brownian motion of particles embedded in a solution of giant micelles. Eur Phys J E 8:431–436
7. Mason TG (2000) Estimating the viscoelastic moduli of complex fluids using the generalized Stokes-Einstein equation. Rheol Acta 39:371–378
8. Faers M (2002) Controlling the state of dispersion and sedimentation stability of colloidal suspensions with both absorbing and non-absorbing polymers. Cahier de Formulation 10:207–222
9. Poon WCK et al (1999) Delayed sedimentation of transient gels in colloid-polymer mixtures: dark field observation, rheology and dynamic light scattering studies. Faraday Discuss 112:143–154

Progr Colloid Polym Sci (2012) 139
DOI: 10.1007/978-3-642-28974-3_18

Microfiltration of Deforming Droplets

A. Ullah[1,2], M. Naeem[3], R.G. Holdich[1], V.M. Starov[1], and S. Semenov[1]

Abstract Control of permeate flux is important in microfiltration processes as it influences trans-membrane pressure and fouling of a membrane. Particles of vegetable oil ranging from 1 to 15 μm were passed through a 4 μm slotted pore membrane at various flux rates. Various intensities of shear were applied parallel to the membrane by vibrating the membrane at different frequencies. At the lowest permeate flux rate (200 $\mathrm{l\,m^{-2}\,hr^{-1}}$) the membrane fouled because the drag force was too low to squeeze the deformable oil droplets through the membrane. At higher flux rates the drag force over the oil droplets increased and deformation, and passage, of oil droplets into the permeate was possible. Without any applied shear highest trans-membrane pressure was observed due to fouling, which could be modelled by a pore blocking model. A positive displacement pump was used in experiments which maintained nearly constant flow of permeate. Flux rates varied from 200 up to 1200 $\mathrm{l\,m^{-2}\,hr^{-1}}$, and the highest shear rate used was 8,000 s^{-1}. The experimental system provided a simple technique for assessing the behaviour of the microfilter during the filtration of these deforming particles.

Introduction

A variety of approaches have been adopted to reduce fouling of a membrane used in microfiltration. For example, fouling can be reduced and permeate flux increased with increasing ionic strength of TiO_2 [1]. Concentration is in other important factor that influences fouling and permeate flux. The flux rate of latex suspensions linearly decreased with increase in its concentration [2]. Another study showed that increase in concentration of bacteria suspension does not foul the membrane [3]. Particle size does severely influence the fouling of membrane. Keeping operating conditions constant the influence of particle size on membrane fouling was studied [4]. It is found that retaining particles on the surface of a membrane decreases the fouling effect, compared to retaining particles inside the membrane [5]. Small membrane pores of are more susceptible to fouling, due to not allowing particles to pass through the membrane [6]. More recently, fouling has been reduced by making a membrane surface hydrophobic, which gave higher permeation flux rates [7]. Higher permeation flux of hydrophobic membranes is due to the super-hydrophobic and super-olephilic nature of oil [8].

A. Ullah (✉)
[1]Department of Chemical Engineering, Loughborough University, Loughborough, Leicestershire, LE11 3TU, UK
e-mail: A.Ullah@lboro.ac.uk
[2]Department of Chemical Engineering, NWFP (KPK), UET, Peshawar, Pakistan
[3]Department of Chemistry, AWKUM, Mardan, Pakistan

Theory

There are a number of equations that may be used to correlate permeate flow rate and the required pressure drop through a membrane: for example Hagen-Poiseuille and Darcy's law. For laminar flow they simply state that the pressure drop required for the flow increases linearly with the permeate flow rate. In all cases there is a constant of proportionality that is a resistance term, or some form of inverse permeability, for example:

$$J = \frac{k_o}{\frac{1}{\Delta P} + R} \tag{1}$$

Where J is the permeate flux through the membrane, $k_o$is the initial permeability of membrane, ΔP is the pressure difference across the membrane and R is resistance to flow from the membrane and its deposit. Assuming a blocking model of filtration [9], then R will change with the amount of material deposited on the membrane as follows:

$$R = k_o c y t$$

Where c is the concentration of dispersion in the permeate, y is the blocking constant and t is filtration time.

Hence, the filtration pressure is related to the flux and time by:

$$\Delta P = \frac{J}{k_o (1 - cyJt)} \quad (2)$$

Resistance "R" to the flow of fluid is dependent on concentration of particles (i.e. oil) in the emulsion, blocking constant of the membrane, permeate flux rate and time. Higher concentrations of oil would offer higher membrane resistance to the flow of fluid. The blocking constant can be related to the physical properties of the membrane and the particles being filtered [9]:

$$y = (\pi \beta d^{**}/4) \int_{d^{**}}^{D\max} f_p(D) dD \quad (3)$$

where $f_p(D)$ is probability distribution function of particle diameter, $D_{\max}$ is maximum particle diameter and β is the hydrodynamic specific surface force of interaction between the surface of the membrane pores/slots and the particles [9]. The constant d^{**} is assumed to be the diameter of pore/slot at the start of the process when it is assumed to be completely open, D is the diameter of particles/droplets that will cause fouling (between d^{**} and the maximum dispersed phase size). Hence, y is the blocking constant and independent of time for any given set of filtration conditions. The model presented in (2) can be applied to circular and slotted pore membranes with filtration of non-deforming particles. It may be valid up to a point when filtering deforming oil drops; i.e. where the passing of deformed drops through the membrane may result in higher flux rates, or lower operating pressures, than would otherwise be obtained when filtering solid (non-deforming) particles. In addition, the model is one that applies the full concentration of the dispersed phase (c) and blocking constant (y) to the analysis. During conditions of microfiltration applying shear at the membrane surface, rather than dead-end filtration, then it could be argued that either (or both) of these constants may be altered compared to those obtained in the absence of shear.

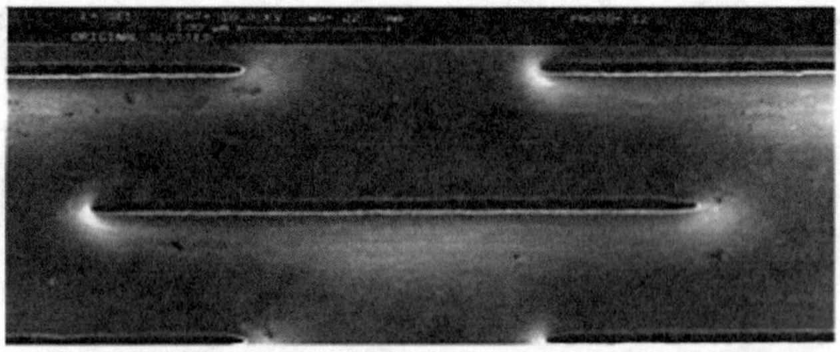

Fig. 1 Image of the surface of a slotted pore membrane

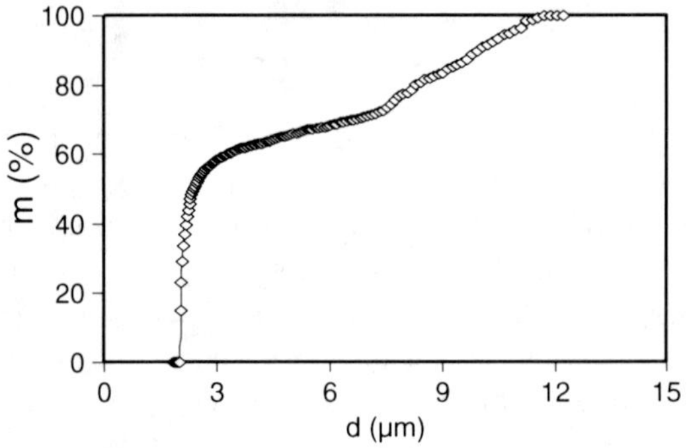

Fig. 2 Typical size distribution (cumulative mass undersized "m" versus droplet diameter "d") of the oil droplets produced

Experimental

Material

The oil used was vegetable oil from a local supermarket (EU Rapeseed from the Co-operative Group Ltd, UK). Silica (SiO_2) (Degussa AG, Germany) was used to enhance oil droplet stability by decreasing the deformation of the droplets when subjected to high trans-membrane pressure during filtration, by increasing the interfacial tension between the oil droplet and water. Permeate flux was sucked through the membrane by a peristaltic pump (RS 440–515, UK). To measure trans-membrane pressure a sensitive pressure transducer (HCX001A60, Farnell, UK) was used. Filtration experiments were carried out with a 4 μm slotted pore size (Micropore Technologies Ltd, UK). This is an unusual membrane filter design, as the pore is in the form of a slot and it has no internal tortuosity: it is similar to an industrial screen, but with a slot width of 4 μm and a slot length of 400 μm, see Fig. 1. Shear rates were applied on the surface of membrane and membrane frequency was controlled through a voltage controller (Deltra eteckronika 1464). For droplet size distribution and concentration determination a Coulter Multisier II (Coulter Counter, Coulter Electronics Ltd) was used. A Coulter was used so that the concentration of droplets within each size grade, or range, could be measured and it was possible to deduce the full 'grade efficiency' curve for each filtration.

Filtration

One to fifteen micrometre vegetable oil particles were produced with a food blender operated using its highest speed for 12 min, a typical size distribution is illustrated in Fig. 2. All filtration experiments were carried out using a 4 μm

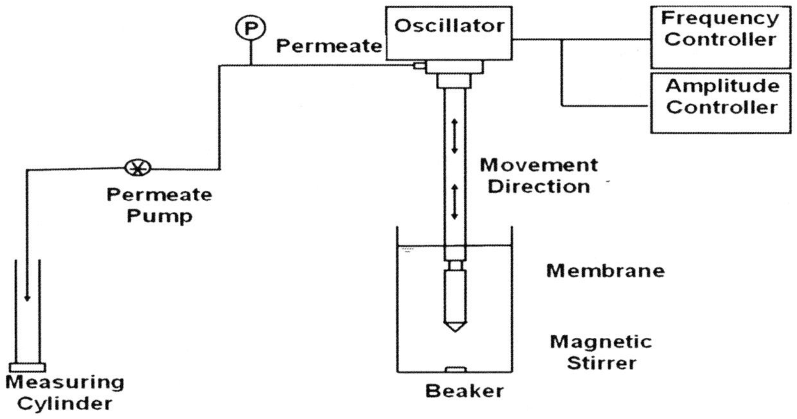

Fig. 3 Schematic view of vibrating microfiltration rig

slotted pore membrane attached to a vibrating arm activated by an electromechanical oscillator. A replaceable membrane was attached to the vibrating head using a hollow steel rod that provided both structural support as well as the permeate flow tube. Frequency and amplitude of the membrane were controlled and were adjusted between 0–100 Hz and 0–10 mm, respectively. Vibration is produced in the vertical direction that creates shear on the outer surface of the membrane. Emulsion feed was contained outside of the membrane and permeate was sucked, by a peristaltic pump (a positive displacement type of pump enabling almost constant flow rate conditions), and periodically collected in a measuring cylinder, trans-membrane pressure was recorded and the permeate returned to the feed tank. Trans-membrane pressure is defined here as the difference in pressures between the inside of the membrane and atmospheric pressure.

Filtration of vegetable/water oil emulsion was carried out with, and with-out, vibration of the membrane using various frequencies and flux rates. Before and after each run the membrane was cleaned with 2% Ultrasil 11 and hot (50°C) filtered water. At different trans-membrane pressures various flux rates were obtained and compared with clean water flux rates at the respective trans-membrane pressures. Equilibrium conditions are those when trans-membrane pressures and flux rates do not change with time. The membrane was vibrated vertically with various frequencies and, due to it, different intensities of shear rates were applied to the membrane. The effect of different intensities of shear rates were studied at the different flux rates using vibrating microfiltration rig see Fig. 3.

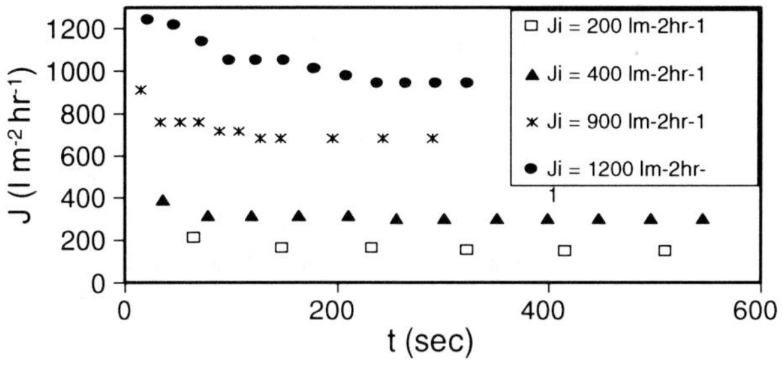

Fig. 4 Experimental measurements of flux rates with time during the oscillating filtration experiments at 0 s^{-1} shear rate

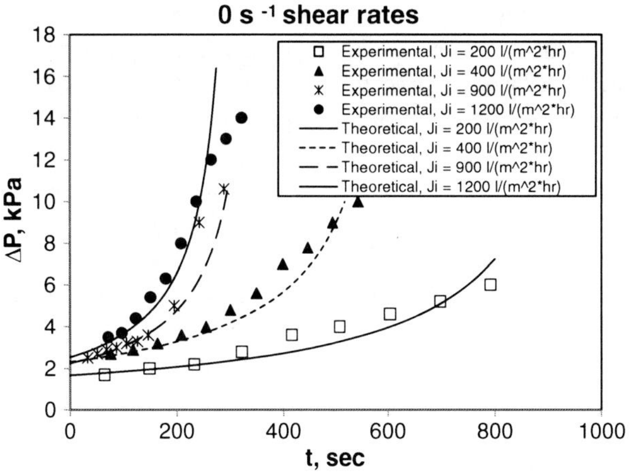

Fig. 5 Experimental measurements and theoretical predictions of pressure drop with time during the oscillating filtration experiments using the blocking model at 0 s^{-1} shear rate

Results and Discussions

In all filtrations nearly constant flux was achieved, due to the reason of using positive displacement pump. When membrane fouling occurred, the permeate flux rate did not significantly decline, instead it was the trans-membrane pressure that increased, see Figs. 4 and 5.

Higher trans-membrane pressure was observed in the absence of any surface shear (vibrations). A blocking model for the microfilter appears to describe the performance adequately, at least up until a threshold such as at 250 s for the 1,200 $l m^{-2} h^{-1}$ permeate flux curve, see Fig. 5.

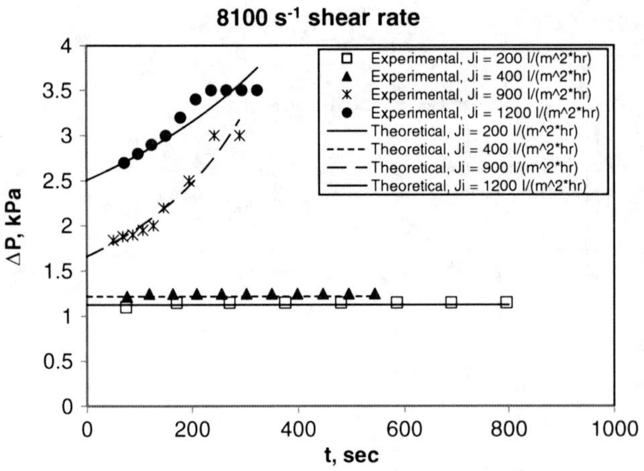

Fig. 6 Experimental measurements and theoretical predictions of pressure drop with time during the oscillating filtration experiments using the blocking model at 8,100 s^{-1} shear rate applied

A considerable change in trans-membrane was observed with applying shear rate (8,000 s^{-1}) to the membrane. Shear decreases fouling of the membrane, as can be seen by the lower trans-membrane pressure obtained without shear illustrated by comparing Fig. 6 with Fig. 5.

Conclusions

Trans-membrane pressure can be affected by applying shear rate. Without shear higher trans-membrane pressure was observed due to higher fouling of the membrane. At lower flux rates deformation of oil droplets was lower and mostly the droplets were retained on the surface of the membrane. Under these conditions the trans-membrane pressure increased linearly with time, in a conventional way that may be expected during 'constant rate' filtration. However, the concentration of material being filtered was low, hence the increase in pressure is due to increased blocking at the membrane surface, and not by the formation of a filter cake. After a certain pressure drop is obtained, equating to an increase in the local velocity of liquid flowing through the pores that remain open on the membrane filter, then the oil drops that can deform are squeezed through the pores from the feed vessel and into the permeate. The blocking model is no longer applicable under these conditions, as some of the dispersed phase is now being sucked into the permeate and no longer contributing towards the membrane blocking. Applying shear to the membrane surface reduces the membrane blocking and, therefore, the trans-membrane pressure was lower. The reduction in blocking could be correlated with increasing applied shear.

References

1. Zhao Y, Xing W, Xu N, Wong FS (2005) Effects of inorganic salt on ceramic membrane microfiltration of titanium dioxide suspension. J Membr Sci 254:81–88
2. Guiziou GG, Wakeman RJ, Daufin G (2002) Stability of latex crossflow filtration: cake properties and critical conditions of deposition. Chem Eng J 85:27–34
3. Persson A, Jonsson AS, Zacchi G (2001) Separation of lactic acid producing bacteria from fermentation broth using a ceramic microfiltration membrane with constant permeate flow. Biotechnol Bioeng 72:269–277
4. Kwon DY, Vigneswaran S, Fane AG, Aim RB (2000) Experimental determination of critical flux in cross-flow microfiltration. Sep Purif Technol 19:169–181
5. Holdich RG, Kosvintsev S, Cumming IW, Zhdanov S (2006) Pore design and engineering for filters and membrane. Phil Trans R Soc A 364:161–174
6. Cumming IW, Holdich RG, Smith ID (2000) The rejection of oil microfiltration of a stabilised kerosene/water emulsion. J Membr Sci 169:147–155
7. Lee C, Baik S (2010) Vertically-aligned carbon nano-tube membrane filters with superhydrophobicity and superoleophilicity. Carbon 48:2192–2197
8. Lee CH, Johnson N, Drelich J, Yap YK (2011) The performance of superhydrophobic and superoleophilic carbon nanotube meshes in water–oil filtration. Carbon 49:669–676
9. Filippov A, Starov VM, Lloyd RD, Chakravarti S, Glaser S (1994) Sieve mechanism of microfiltration. J Membr Sci 89:199–213

Progr Colloid Polym Sci (2012) 139
DOI: 10.1007/978-3-642-28974-3_19

Fabrication of Biodegradable Poly(Lactic Acid) Particles in Flow-Focusing Glass Capillary Devices

Goran T. Vladisavljević[1,2], J.V. Henry[1], Wynter J. Duncanson[3], Ho C. Shum[4], and David A. Weitz[3]

Abstract Monodisperse poly(dl-lactic acid) (PLA) particles with a diameter in the range from 12 to 100 μm were fabricated in flow focusing glass capillary devices by evaporation of dichloromethane (DCM) from emulsions at room temperature. The dispersed phase was 5% (w/w) PLA in DCM containing a small amount of Nile red and the continuous phase was 5% (w/w) poly(vinyl alcohol) in reverse osmosis water. Particle diameter was 2.7 times smaller than the size of the emulsion droplet template indicating that the particle porosity was very low. SEM images revealed that the majority of particle pores are in the sub-micron region but in some instances these pores can reach 3 μm in diameter. Droplet diameter was influenced by the flow rates of the two phases and the entry diameter of the collection capillary tube; droplet diameters decreased with increasing values of the flow rate ratio of the dispersed to continuous phase to reach constant minimum values at 40–60% orifice diameter. At flow rate ratios less than 5, jetting can occur, giving rise to large droplets formed by detachment from relatively long jets (~10 times longer than droplet diameter).

Introduction

Microspheres composed of biodegradable polymers have been used for the encapsulation and controlled release of hydrophilic and hydrophobic drugs [1], ultrasound and molecular imaging [2], cell cultivation in tissue engineering [3], fabrication of scaffolds for bone tissue repair applications [4], fabrication of composite coatings for implantable devices [5], etc. The most commonly used biodegradable synthetic polymers for these applications are poly(lactic acid), PLA, and poly(lactic-co-glycolic) acid, PLGA, since they both have favourable properties such as good biocompatibility, biodegradability, and mechanical strength [1, 6].

PLA particles are usually produced via single emulsion route shown in Fig. 1a. In this process, a mixture of PLA and volatile organic solvent is emulsified in the aqueous surfactant solution. The second step involves solvent evaporation or extraction, resulting in shrinkage of the droplets and formation of a coherent solid particle as shown in Fig. 1b. These particles are suitable for encapsulation of hydrophobic drugs [7].

The PLA droplets shrank during solvent (DCM) evaporation, as demonstrated in Fig. 1b. In this figure the average emulsion droplet diameter was 64 μm and the resulting particle diameter (at the right-hand edge of Fig. 1b) was 24 μm, which agrees well with theoretical value of 64/2.7 = 24 μm calculated on the assumption of complete solvent evaporation and zero particle porosity.

Precision generation of drops is a crucial step in fabrication of PLA particles because monodisperse particles can only be produced from monodisperse droplets. Conventional methods for production of PLA particles based on spray drying [8] or homogenization in a stirred vessel or high-pressure homogenizers result in polydisperse particles. For example, Straub et al. [8] have produced PLGA particles with a mean size of 2.3 μm using a spray dryer at 20 mL min^{-1} emulsion flow rate, but the particle size in the product was in the range between 1 and 10 μm. Monodisperse particles are favourable in drug delivery and ultrasound imaging applications, because particles behave homogenously, ensuring a predictable release rate profile and acoustic response. Several drop-by-drop methods have been used in production of PLA and PLGA microspheres, such as ink-jet printing [9], direct and pre-mix membrane

G.T. Vladisavljević (✉)
[1]Department of Chemical Engineering, Loughborough University, Loughborough, LE11 3TU, UK
e-mail: G.Vladisavljevic@lboro.ac.uk
[2]Vinca Institute of Nuclear Sciences, University of Belgrade, 522, Belgrade, Serbia
[3]Department of Physics, Harvard University, Cambridge, MA 02138, USA
[4]Department of Mechanical Engineering, University of Hong Kong, Hong Kong, China

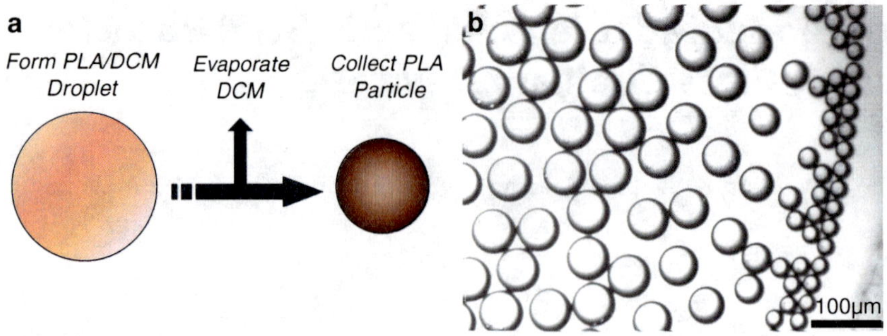

Fig. 1 Emulsion templating method for producing PLA microparticles. (**a**) Schematic diagram of process; (**b**) micrograph of emulsions undergoing DCM evaporation

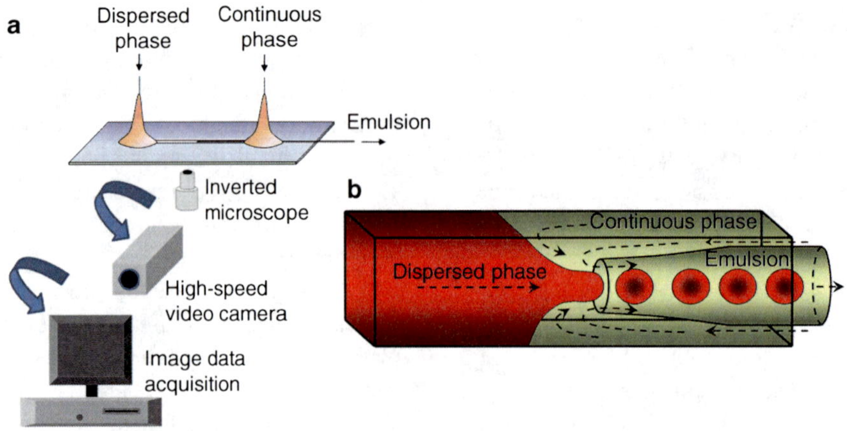

Fig. 2 Experimental setup for the two-phase flow in a flow focusing configuration. (**a**) Visual representation of data collection; (**b**) enlarged view of flow focusing zone in device, where droplets are formed

emulsification [7–10], extrusion through a nozzle under acoustic excitation [11] and flow focusing in PDMS microfluidic devices [12]. However, these methods either produce particles with high coefficient of variations or are expensive to build and to operate. For example, coefficient of variations of particle sizes for PLGA particles produced by direct membrane emulsification was 7–16% [8]. Using repeated pre-mix membrane emulsification, the mean particle size of around 1 μm was achieved but the span of size distribution was 0.7 [10]. The purpose of this study is to fabricate inexpensive glass capillary devices, developed by Utada et al. [13], shown in Fig. 2, so that highly uniform PLA microspheres may be produced economically.

Materials and Methods

Glass capillaries of inner diameter 580 μm and outer diameter 1 mm were pulled using a Sutter Instruments model P-97 micropipette puller to obtain a tapered tip. Each tip was sanded smooth to obtain an orifice with an entry diameter of 60–280 μm, followed by surface modification with 2-[methoxy (polyethylenoxy)propyl]trimethoxysilane (Gelest Inc.). These capillaries with hydrophilic coating were then inserted and glued in the centre of a square glass tube (AIT Inc.) of 1.05 mm inner dimension. The oil phase was 5% (w/w) poly (dl-lactic acid) (Polysciences Europe GmbH, molecular weight=15,000 g mol^{-1}) in dichloromethane (DCM) containing small amount of Nile red dye (Sigma-Aldrich). The role of dye in our preparation was to easily observe the position of the interface between the organic and aqueous phase in the device because Nile red is insoluble in water. The viscosity and density of the organic phase is 0.3631 mPa·s and 1.31 kg·dm^{-3} respectively at 298 K. The aqueous phase was 5% (w/w) polyvinyl alcohol (Sigma-Aldrich, 87–89% hydrolyzed) in Milli-Q water. The interfacial tension between the two phases was measured as 2.26 mN/m using a Krüss DSA-100 pendant drop tensiometer. The both phases were delivered to the device from gas tight syringes using separate Harvard Apparatus PHD 22/2000 syringe pumps connected to 19-gauge needles, as shown in Fig. 2a.

Flow optimization (through manually reprogramming flow rates of both phases) was enabled by real-time visualization of emulsion formation using Phantom V5.1 high-speed camera focused by an inverted microscope. The images are obtained using high-speed video footage at 1,000–2,000 frames per second. Droplet diameters were

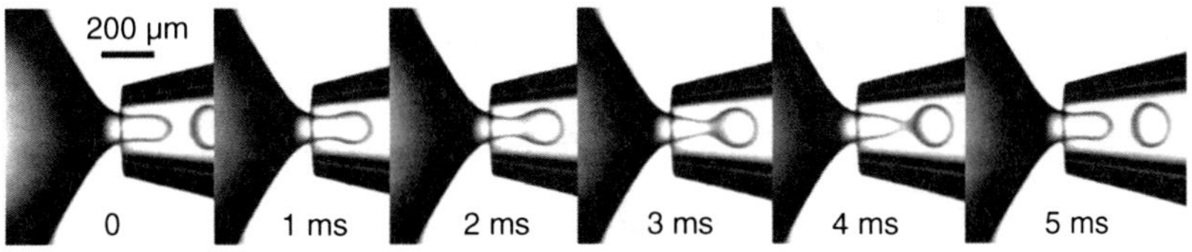

Fig. 3 Six consecutive frames of droplet production via dripping mechanism over the course of 5 ms. The drop generation frequency was 206 Hz

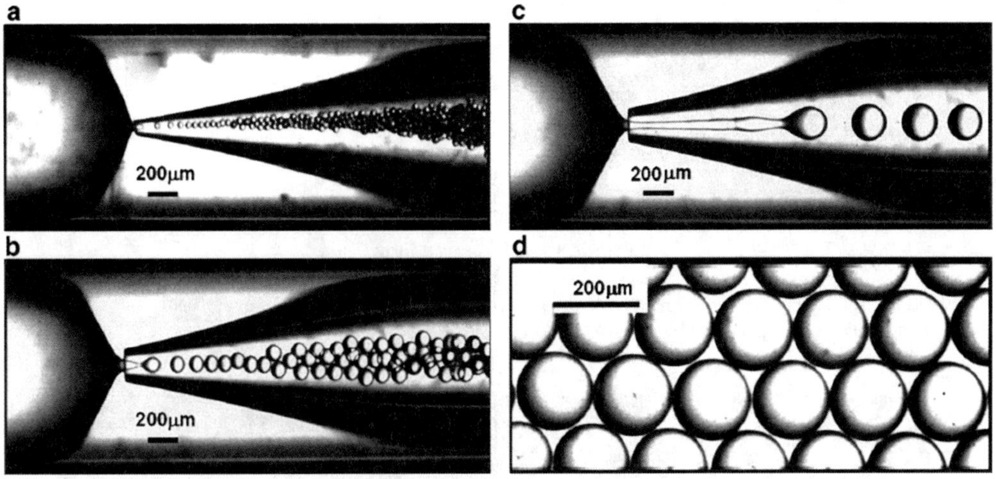

Fig. 4 Images of droplet formation under varying flow rates and orifice size. (**a**) Q_c = 0.5 ml·h^{-1}, Q_d = 0.003 ml·h^{-1}, $d_{orifice}$=60 μm, d = 33 μm; (**b**) Q_c = 5 ml·h^{-1}, Q_d = 1 ml·h^{-1}, $d_{orifice}$=130 μm, d = 100 μm; (**c**) Q_c=6.5 ml·h^{-1}, Q_d=0.7 ml·h^{-1}, $d_{orifice}$=130 μm, d = 230 μm; (**d**) collected monodisperse droplets

measured using ImageJ v.1.44 software. Frames of the high-speed video footage were taken to allow individual measurement of the droplet diameters. Using this technique also, stills were studied to characterize the drop generation behavior and estimate drop generation rate.

Examples of the video footage gathered at 1,000 fps are shown in Fig. 3. In this experiment, the flow rate of the dispersed phase, Q_d was set at 1.5 ml·h^{-1}, the flow rate of the continuous phase, Q_c, was set at 13 ml·h^{-1} and the resulting observed diameter of the droplets was d= 156 μm. The following mass balance equation can be used for the dispersed phase: $f = Q_d/(d^3\pi/6)$, where f is the frequency of drop generation. Putting $d = 1.56 \times 10^{-4}$ m and $Q_d = 4.17 \times 10^{-10}$ m^3/s into the above equation, one obtains 210 Hz which is in good agreement with 206 Hz estimated from the video recordings.

The fluid flow rates and device geometry has an important effect on droplet size, as illustrated by the examples in Fig. 4. In (a), the orifice size was made very small so as to obtain relatively small droplets with a diameter of 33 μm that can be converted into 12 μm particles after solvent evaporation. In (b) and (c) the device was the same but the flow rates were different, leading to different droplet formation regimes. As can be seen in (d) collected droplets are packed into regular hexagonal arrangements, evidence of a high degree of monodispersity, with a coefficient of variation of less than 2%.

In Fig. 5, the relationship between the dimensionless droplet diameter, $d/d_{orifice}$ and flow rate ratio, Q_c/Q_d is shown. There are two different regimes of droplet formation, e.g. jetting and dripping. At low flow rate ratios, where $Q_c/Q_d < 5$ jetting can occur, which gives rise to large droplets and relatively long length jets prior to droplet detachment. At jetting, jet lengths are ~10 times longer than droplet diameter and the droplet diameter can be up to twice the diameter of the orifice. In addition, droplets formed under jetting regime are more polydisperse than droplets formed under dripping regime. These regimes are commutative at $Q_c/Q_d < 5$, and any small disturbance received by the system may either cause or destroy long jetting breakup. The second feature of Fig. 5 is that droplet diameters decrease with increasing values of Q_c/Q_d to reach their respective minimum values at 40–60% orifice diameter. At flow rate ratios between 5 and 15, droplets have the highest degree of monodispersity. At $Q_c/Q_d > 15$, a second region of droplet formation from extended jetting predominates; these extensional flows however, are not as long as those encountered at $Q_c/Q_d < 5$, nor as wide as those exhibited at these values; for this reason are not subject to the same instability, but instead exhibit a tip streaming regime, which is reflected in the smaller, more polydisperse droplet diameters.

The polylactic acid particles have a high degree of monodispersity and low degrees of attrition and porosity, as shown in the optical and electron microscope images in

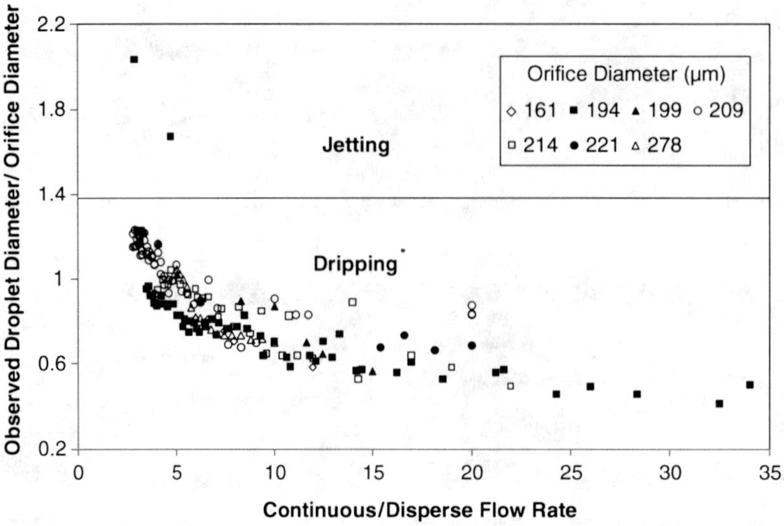

Fig. 5 Graph of droplet diameter/orifice diameter versus flow rate ratio, Q_c/Q_d, for a range of device internal orifice diameters: ◇ = 161 μm; ■ = 194 μm; ▲ = 199 μm; ○ = 209 μm; □ = 214 μm; ● = 221 μm; Δ = 278 μm

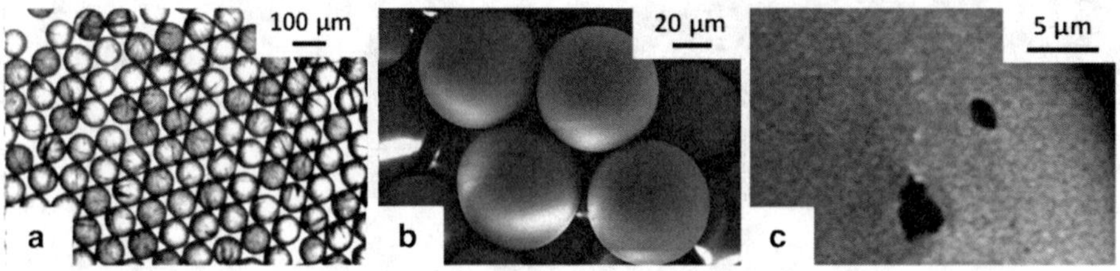

Fig. 6 (**a**) Optical micrograph of 84 μm poly(lactic) acid particles; (**b**) scanning electron micrograph of 63 μm particles; (**c**) surface of a PLA particle showing two micrometer-sized pores

Fig. 6. The coefficient of variation of particle sizes in Fig. 6a is 2.3%. Examination of SEM images has revealed that the surface of PLA particles was essentially non-porous with a majority of pores in the sub-micron region but in some rare instances these pores can reach 3 μm in diameter, as shown in Fig. 6c.

Conclusions

The diameter of PLA/DCM droplets in flow focusing glass capillary devices has been closely controlled by phase flow rates and entry diameter of collection capillary tube. The maximum emulsion and particle monodispersity was achieved at flow rate ratios of continuous to dispersed phase between 5 and 15; outside of this range, jetting during the formation of emulsion droplets predominates, and is accompanied by more polydisperse and for lower flow rate ratios, larger droplet diameters. Particle diameters have been established to be 2.7 times smaller than the emulsion diameters. The resultant particles have low porosity, and are highly monodisperse.

Acknowledgment The work was supported by the Engineering and Physical Sciences Research Council (EPSRC) of the United Kingdom (grant reference number: EP/HO29923/1).

References

1. Jain JA (2000) Biomaterials 21:2475
2. Sawalha H, Schroën K, Boom R (2011) Chem Eng J 169:1
3. Klibanov AL (2007) J Nucl Cardiol 14:876
4. Shi X, Sun L, Jiang J, Zhang X, Ding W, Gan Z (2009) Macromol Biosci 9:1211
5. Wang Y, Shi X, Ren L, Wang C, Wang DA (2009) Mater Sci Eng C 29:2502
6. Bhardwaj U, Papadimitrakopoulos F, Burgess DJ (2008) J Diabetes Sci Technol 2:1016
7. Ito F, Makino K (2004) Colloid Surf B 39:17
8. Straub JA, Chickering DE, Church CC, Shah B, Hanlon T, Bernstein H (2005) J Control Release 108:21
9. Böhmer MR, Schroeders R, Steenbakkers JAM, de Winter SHPM, Duineveld PA, Lub J, Nijssen WPM, Pikkemaat JA, Stapert HR (2006) Colloid Surf A 289:96
10. Sawalha H, Purwanti N, Rinzema A, Schroën K, Boom R (2008) J Membr Sci 310:484
11. Berkland C, Kim KK, Pack DW (2001) J Control Release 73:59
12. Xu Q, Hashimoto M, Dang TT, Hoare T, Kohane DS, Whitesides GM, Langer R, Anderson DG (2009) Small 5:1575
13. Utada AS, Chu L-Y, Fernandez-Nieves A, Link DR, Holtze C, Weitz DA (2007) MRS Bull 32:702

Progr Colloid Polym Sci (2012) 139
DOI: 10.1007/978-3-642-28974-3_20

Control over the Shell Thickness of Core/Shell Drops in Three-Phase Glass Capillary Devices

Goran T. Vladisavljević[1,2], Ho Cheung Shum[3], and David A. Weitz[4]

Abstract Monodisperse core/shell drops with aqueous core and poly(dimethylsiloxane) (PDMS) shell of controllable thickness have been produced using a glass microcapillary device that combines co-flow and flow-focusing geometries. The throughput of the droplet generation was high, with droplet generation frequency in the range from 1,000 to 10,000 Hz. The size of the droplets can be tuned by changing the flow rate of the continuous phase. The technique enables control over the shell thickness through adjusting the flow rate ratio of the middle to inner phase. As the flow rate of the middle and inner phase increases, the droplet breakup occurs in the dripping-to-jetting transition regime, with each double emulsion droplet containing two monodisperse internal aqueous droplets. The resultant drops can be used subsequently as templates for monodisperse polymer capsules with a single or multiple inner compartments, as well as functional vesicles such as liposomes, polymersomes and colloidosomes.

Introduction

Droplets with a core/shell structure can be used as templates for the production of hollow particles, such as poly(lactic acid) microbubbles [1], hydrogel shells [2] and liquid crystal shells [3], as well as vesicles, such as liposomes [4], polymersomes [5] and colloidosomes [6]. Conventional techniques for vesicle preparation (e.g. rehydration of dried amphiphilic films and electroforming) rely on the self-assembly of amphiphiles under shear and electric field, respectively [7]. However, the resultant vesicles are often non-uniform in size and structure due to the random nature of the self-assembly process. In addition, encapsulation yields of active materials are relatively low [8]. With microfluidic core/shell droplets as templates, vesicles can be prepared via directed assembly of the amphiphiles with higher degree of control over the droplet size and size uniformity as well as improved encapsulation efficiency. Regarding other core/shell particles, conventional techniques include internal phase separation [9], complex coacervation [10], layer-by-layer deposition [11] and interfacial polymerization [12]; these typically require multi-stage processing, very specific emulsion formulation, and does not allow close control over the shell thickness. The control over the shell thickness of core/shell particles is highly relevant, because it affects their robustness, mechanical properties and permeability to different species. Poly(dimethylsiloxane) (PDMS) microfluidic devices containing sequential junctions with alternating wettability have been used to generate controllable multiple emulsions [13]. However, PDMS swells in strong organic solvents and has inherently unstable surface properties. In addition, due to its high elasticity, PDMS channels tend to deform with applied pressure. In this work, monodisperse multiple emulsion droplets of different morphology (core/shell or with two inner droplets per each outer drop) have been produced using glass capillary microfluidics [14, 15]. Glass is more chemically robust than PDMS, does not swell, and can easily be functionalized to control surface properties. The aim of the research was to investigate the effect of fluid flow rates on droplet size, morphology, and shell thickness.

Experimental

Controllable multiple emulsions were produced using microcapillary device shown in Fig. 1. The device consists of two round capillaries with a tapered end inserted into a

G.T. Vladisavljević (✉)
[1]Department of Chemical Engineering, Loughborough University, Loughborough LE11 3TU, UK
e-mail: G.Vladisavljevic@lboro.ac.uk
[2]Vinča Institute of Nuclear Sciences, PO Box 522, Belgrade, Serbia
[3]Department of Mechanical Engineering, University of Hong Kong, Hong Kong, China
[4]Department of Physics, Harvard University, Cambridge, MA 02138, USA

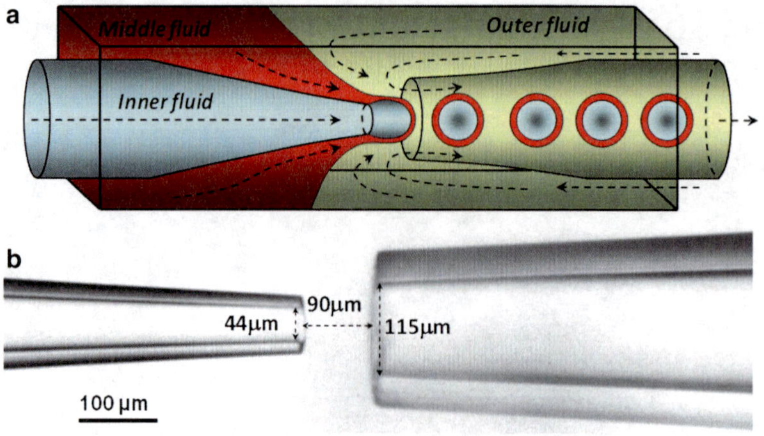

Fig. 1 (**a**) Schematic of the glass capillary device used in this work; (**b**) micrograph of the tapered end of the injection and collection tube. The diameter of the orifice in the injection and collection tube was 44 and 115 μm, respectively and the distance between the ends of the two capillaries was 90 μm

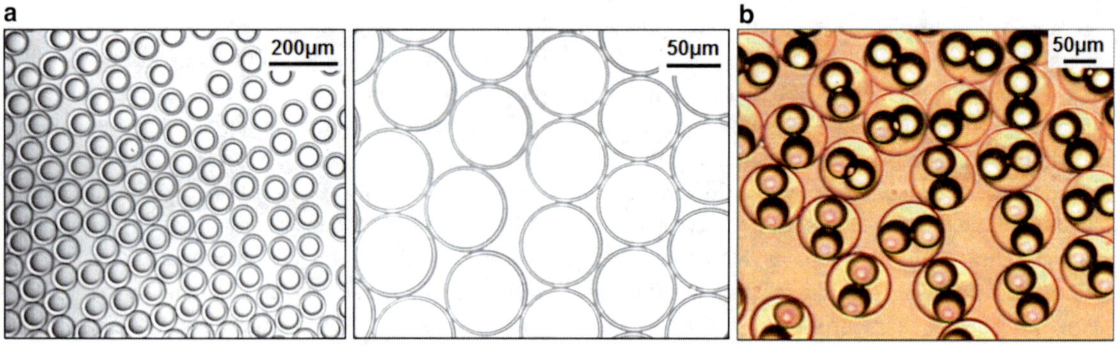

Fig. 2 (**a**) Micrographs of core/shell droplets with d_o=78 μm and δ=8 μm taken under different magnification; (**b**) micrograph of multiple emulsion droplets containing two inner drops

square capillary. First, a round glass capillary (World Precision Instrument, Inc., Sarasota, FL) with an outer diameter of 1 mm and in inner diameter of 0.58 mm was heated and pulled using a Sutter Instrument model P-97 micropipette puller. As a result of the pulling process, the capillary breaks into two parts, each with a tapered end that culminates in a fine orifice. The orifice in both capillaries was then enlarged to the desired diameter by microforging using a Narishige model MF-830 microforge. Capillaries with the desired orifice size were then inserted into a square glass capillary (Vitrocom, Fiber Optic Center, Inc., New Bredford, MA) and glued together on a microscope glass slide. Coaxial alignment of the capillaries was ensured by choosing the capillaries such that the outer diameter of the round capillaries matches the inner dimension of the square capillary. Milli-q water (the inner fluid) was introduced through the injection tube and a solution of 2% (w/w) Dow Corning 749 Fluid in 10 cP PDMS (the middle fluid) was supplied cocurrently through the square capillary. Two percentange (w/w) poly(vinyl alcohol) aqueous solution was supplied from the opposite side through the square capillary and all phases were forced into the collection tube. It resulted in the rupture of coaxial jet composed of the inner and middle fluid and formation of droplets, as shown in Fig. 1a. In order to prevent wetting of the injection tube with water, the tapered region of the injection tube was rendered hydrophobic with octadecyltrimethoxysilane. All liquids were delivered to the device from gas tight syringes using separate Harvard Apparatus PHD 22/2000 syringe pumps interfaced to a PC computer. The drop generation process was observed in real time and recorded using a Phantom v7.0 fast speed camera (Vision Research, Inc., USA) attached to a Leica DMIRB inverted microscope.

Droplet Size and Generation Frequency

In the dripping regime, the interfacial tension between the middle and outer fluid dominates the droplet breakup process [16] and the double emulsions generated had one internal aqueous drop (Fig. 2a). As the inertial force of the middle and inner phases becomes comparable to the interfacial tension, the droplet breakup occurs in the dripping-to-jetting transition regime. Each double emulsion droplet formed in this regime had two monodisperse internal aqueous droplets, as shown in Fig. 2b. Drops with two inner droplets can be used as templates for fabrication of vesicles with multiple inner compartments, for example multi-compartment colloidosomes [17] or alternatively, dimer particles can be produced, if inner droplets are polymerized.

The diameter, d_o and generation frequency, v of the droplets were controlled by changing the flow rate of the continuous phase, Q_o in the range from 10 to 45 mL h^{-1} while keeping the constant flow rate of the middle and inner phase. At Q_i = 4 and Q_m =1 mL h^{-1}, the drop generation frequency was in the range from 1,000 to 10,000 Hz and the droplet diameter varied widely between 60 and 150 μm, as shown in Fig. 3. The droplets in this size range can be useful for subcutaneous drug release [18]. As Q_o increases, the viscous drag force between the outer and middle phase increases causing the droplets to detach sooner from the injection tube. As a result, smaller droplets are produced at higher frequencies. The drop generation frequency was nearly 10,000 Hz at Q_o= 45 mL h^{-1}. Above that threshold value, the middle fluid forms a long jet, which breaks farther downstream. This regime results in the formation of polydisperse droplets because the point at which a drop separates from the jet can vary. A mass balance for the drop generation process in the dripping regime gives:

$$d_o = [6(Q_m + Q_i)/(\pi v)]^{1/3} \quad (1)$$

The drop diameters calculated from (1) were consistent with the experimental d_o values estimated from microscopic images, as shown in Fig. 3. Equation 1 shows that d_o is proportional to $v^{-1/3}$ at constant Q_m and Q_i values, which is confirmed in Fig. 3.

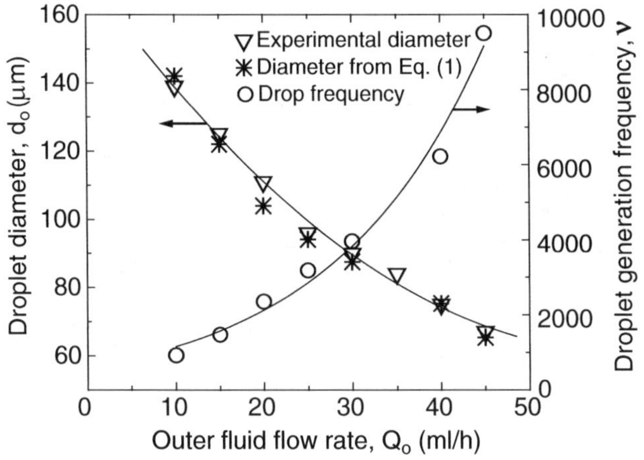

Fig. 3 The effect of outer fluid flow rate on the droplet diameter and generation frequency at Q_i = 4 mL/h and Q_m=1 mL/h. The solid lines represent the best fit of the experimental data

Shell Thickness of Core/Shell Droplets

Assuming that the core is at the center of the shell, which is valid for phases with nearly equal densities, the mass balance equation for middle fluid in the dripping regime gives:

$$\delta = [3Q_i/(4\pi v)]^{1/3}\{[1 + (Q_m/Q_i)]^{1/3} - 1\} \quad (2)$$

where δ is the shell thickness. Equation 2 shows that the shell thickness should increase with increasing the flow rate ratio, Q_m/Q_i, as shown in Fig. 4. At Q_m/Q_i = 6, the middle fluid extends into a long jet before forming a shell around a water droplet. At Q_m/Q_i = 0.25 and 1.5, the drop diameter was greater than the diameter of the orifice in the collection capillary, resulting in deformation of the droplets in the tapered region and formation of ellipsoidal drops. It is visible in the image at Q_m/Q_i = 0.25. If these ellipsoidal drops contain cross-linkable compounds in the shell, they can potentially be transformed into non-spherical particles through polymerisation in the entry region of the collection tube.

For a given flow rate of the middle and inner fluid, the shell thickness decreases with increasing the outer fluid flow rate, as shown in Fig. 5. The minimum shell thickness of

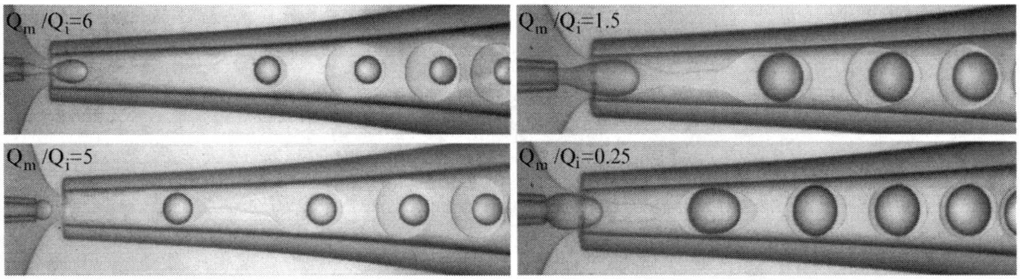

Fig. 4 Still images from high-speed videos showing generation of core/shell droplets with variable shell thickness at different flow rate ratios of the middle to inner phase

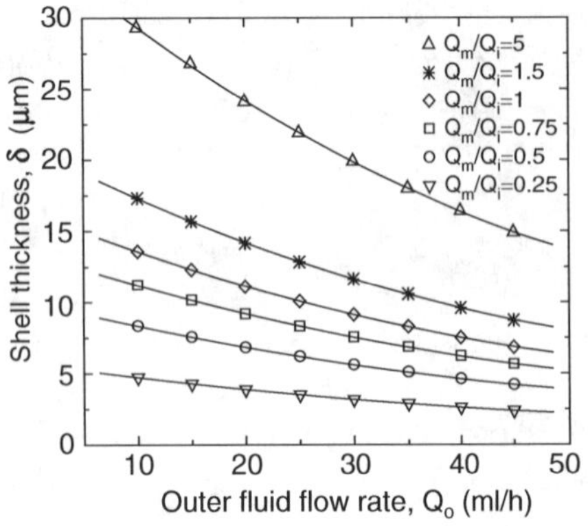

Fig. 5 The shell thickness versus outer fluid flow rate at different flow rate ratios of the inner to middle fluid, Q_m/Q_i. The sum of the middle and inner fluid flow rate was 5 mL h^{-1}

2.4 μm was obtained at the outer fluid flow rate of 45 mL h^{-1} and $Q_m/Q_i = 0.25$. To obtain core/shell droplets with shells in the nanometre region (100 nm or even less), different design of the device should be used with middle fluid delivered through the injection capillary and inner fluid supplied through a small tapered capillary inserted into the injection capillary [19]. In this configuration, the inner and middle fluid form a biphasic flow within the injection capillary and the thickness of the middle fluid can be made very thin because of its high affinity to the wall of the capillary.

Conclusions

We have demonstrated the ability of glass microcapillary device that combine co-flow and flow-focusing geometries to generate controllable multiple emulsions. The droplets formed using this device had either core/shell morphology or contained two internal droplets per each outer drop and the dominant regime was controlled by adjusting fluid flow rates. The shell thickness ranged from about 2 μm to several tens of μm and was finely tuned by controlling the flow rate ratio of the middle to inner fluid, Q_m/Q_i. The higher the Q_m/Q_i ratio, the thicker the shell around the droplet became. The droplet production rate was controlled by adjusting the flow rate of the continuous phase. The shell material can be polymerised to produce spherical or non-spherical polymer shells or may contain dissolved amphiphilic molecules or particles which can undergo self-assembly upon solvent evaporation, leading to the generation of functional vesicles.

Acknowledgement The work was supported by the Engineering and Physical Sciences Research Council (EPSRC) of the United Kingdom (grant reference number: EP/HO29923/1).

References

1. Sawalha H, Schroën K, Boom R (2009) AIChE J 55:2827
2. Kim JW, Utada AS, Fernández-Nieves A, Hu Z, Weitz DA (2007) Angew Chem Int Ed 46:1819
3. Fernández-Nieves A, Vitelli V, Utada AS, Link DR, Márquez M, Nelson DR, Weitz DA (2007) Phys Rev Lett 99:157801
4. Shum HC, Lee D, Yoon I, Kodger T, Weitz DA (2008) Langmuir 24:7651
5. Lorenceau E, Utada AS, Link DR, Cristobal G, Joanicot M, Weitz DA (2005) Langmuir 21:9183
6. Lee D, Weitz DA (2008) Adv Mater 20:3498
7. Walde P, Ichikawa S (2001) Biomol Eng 18:143
8. Shah RK, Shum HC, Rowat AC, Lee D, Agresti JJ, Utada AS, Chu LY, Kim JW, Fernandez-Nieves A, Martinez CJ, Weitz DA (2008) Mater Today 11:18
9. Atkin R, Davies R, Hardy J, Vincent B (2004) Macromolecules 37:7979
10. Nakagawa K, Iwamoto S, Nakajima M, Shono A, Satoh K (2004) J Colloid Interface Sci 278:198
11. Donath E, Sukhorukov GB, Caruso F, Davis SA, Möhwald H (1998) Angew Chem Int Ed 37:2201
12. Han JH, Koo BM, Kim JW, Suh KD (2008) Chem Commun 28:984
13. Abate AR, Weitz DA (2009) Small 5:2030
14. Utada AS, Lorenceau E, Link DR, Kaplan PD, Stone HA, Weitz DA (2005) Science 38:537
15. Sun BJ, Shum HC, Holtze C, Weitz DA (2010) ACS Appl Mater Interfaces 2:3411
16. Utada AS, Fernandez-Nieves A, Stone HA, Weitz DA (2007) Phys Rev Lett 99:4
17. Lee D, Weitz DA (2009) Small 5:1935
18. Gasparini G, Kosvintsev SR, Stillwell MT, Holdich RG (2008) Colloid Surf B 61:199
19. Kim SH, Kim JW, Cho JC, Weitz DA (2011) Lab Chip 11:3162

Index

Printed by Publishers' Graphics LLC